General Preface to the Series

Because it is no longer possible for one textbook to cover the whole field of biology while remaining sufficiently up to date, the Institute of Biology has sponsored this series so that teachers and students can learn about significant developments. The enthusiastic acceptance of 'Studies in Biology' shows that the books are providing authoritative views of biological topics.

The features of the series include the attention given to methods, the selected list of books for further reading and, wherever possible, suggestions for practical work.

Readers' comments will be welcomed by the Education Officer of the Institute.

1979

Institute of Biology
41 Queen's Gate
London SW7 5BR

Preface to the Second Edition

Temperature is a major limiting factor in a wide spectrum of biological processes ranging from the rate of simple chemical reactions to the ecological distribution of an animal species. This booklet is intended to review the various ways in which temperature imposes itself upon the biology of animals and to demonstrate in particular the special advantages possessed by animals which can regulate their deep body temperature.

During the preparation of a Second Edition it was decided not simply to revise and update the First Edition, but to expand it by almost 50%. New material includes a detailed consideration of the central hypothalamic control mechanisms and the biology of fever, while a chapter on Cryobiology has also been added. Although the major emphasis remains on the mammals, further material on poikilothermic forms has been included to maintain the comparative balance of the book.

I should like to express my gratitude to Fiona Hake and to Pauline Doggett for their help with the new additions to the manuscript.

Cambridge, 1978
R. N. H.

Contents

The Institute of Biology's
Studies in Biology no. 35

Temperature and Animal Life

Second Edition

Richard N. Hardy

M.A., Ph.D.
Lecturer in Physiology, University of Cambridge
Fellow and Tutor of Fitzwilliam College

BIOLOGY LAB.

REFERENCE

SOMERSET COLLEGE OF
ARTS AND TECHNOLOGY
DEPT. OF
HUMANITIES
AND SCIENCE

NOT TO BE TAKEN AWAY.

Edward Arnold

© Richard N. Hardy, 1979

First published 1972
by Edward Arnold (Publishers) Limited
41 Bedford Square,
London, WC1B 3DQ

Reprinted 1974
Second Edition 1979

Paper edition ISBN: 0 7131 2752 X

British Library Cataloguing in Publication Data

Hardy, Richard Neville
   Temperature and animal life. – 2nd ed. – (Institute
   of Biology Studies in biology; no. 35
   ISSN 0537-9024).
   1.   Body temperature – Regulation
   I. Title II. Series
   574.1'9'12   QP135
   ISBN 0-7131-2752-X

Printed and bound in Great Britain at
The Camelot Press Ltd, Southampton

1 Introduction

1.1 Temperature and the rate of chemical reactions

1.1.1 The Q_{10} relation

The rate of all chemical reactions, cellular or otherwise, is dependent on temperature. The relationship can be expressed most simply in terms of a temperature coefficient, the Q_{10} value, derived from the equation

$$Q_{10} = \left(\frac{k_1}{k_2}\right)^{10/(t_1 - t_2)}$$

where k_1 and k_2 are the velocity constants (proportional to the rates of reaction) found at temperatures t_1 and t_2 respectively. The Q_{10} for a particular reaction is thus an expression of the predicted increase in rate for a 10°C increase in temperature. For most biological reactions the Q_{10} is between 2 and 3: a Q_{10} of 2.5 indicates an increase in reaction rate of 9.6% per °C increase in temperature.

Unfortunately, in biological systems the value of the Q_{10} itself varies with temperature. For since most metabolic processes virtually cease as the temperature approaches 0°C, the Q_{10} at such temperatures is often high, while as the upper ranges of temperature tolerance are approached, the Q_{10} drops.

1.1.2 The Arrhenius relation

The Q_{10} relationship cannot be explained on the simple basis of increased molecular agitation at higher temperatures, as a 10°C rise in temperature would only cause a 2% increase in the frequency of molecular collisions. However, an explanation for the Q_{10} relationship and a more precise formulation of the influence of temperature on reaction rates can be obtained from the Arrhenius equation:

$$k = Ae^{-Ea/RT}$$

where k = velocity constant;
A = constant relating to molecular collision frequency;
Ea = activation energy (see below);
R = gas constant 8.30 J mol^{-1} K^{-1} (1.98 cal mol^{-1} K^{-1});
T = absolute temperature.

This relationship was first derived empirically by Arrhenius who later explained it by postulating that an 'activated' (high energy) complex lay on the pathway from reactants to products. Under these circumstances, he argued, only those colliding molecules whose combined energy was in excess of the 'activation energy' (Ea) necessary to form the complex would be able to react (Fig. 1–1).

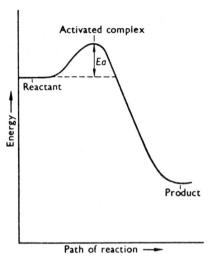

Fig. 1–1   Diagram showing the postulated energy changes during a reaction.

If this is the case, the effect of an increase in temperature can easily be explained. Let us suppose that in a population of colliding molecules at temperature $T°$ (energy distribution being Gaussian) (Fig. 1–2), only a small proportion (diagonal hatching) have a combined energy sufficient to result in a reaction. If the temperature is increased to $(T+x)°$, the energy distribution curve shifts to the right and thus the number of molecules with a combined energy exceeding Ea increases (stippled area). The number of molecules of combined energy exceeding Ea rises in an exponential way as the temperature increases and hence the reaction rate increases exponentially with temperature, as originally shown in the Arrhenius equation.

The Arrhenius equation can be written in logarithmic form:

$$\log\ k = \frac{-Ea}{RT} + C_1$$

$$\log_{10} k = \frac{-Ea}{2.303R} \cdot \frac{1}{T} + C_2$$

where C_1 and C_2 are constants

Hence a plot of log k against $\frac{1}{T}$ will give a straight line from whose gradient Ea can be found (slope $= -\dfrac{-Ea}{2.303R}$).

The activation energy (Ea) is measured in joules or calories and is sometimes known as the 'thermal increment', the 'temperature characteristic' or the 'Arrhenius constant' and is in this case often signified by μ. Its value may give an indication of the nature of the rate-limiting enzyme in a complex sequence of reactions.

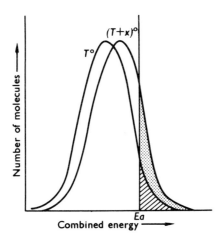

Fig. 1–2   Distribution energy in two populations of molecules at different temperatures.

1.1.3   Enzyme-catalyzed reactions

Reactions which rely upon the presence of enzymes are, like all other reactions, dependent upon temperature, as discussed above. However, in the case of enzymes, another temperature-related factor must be considered: *thermal inactivation*. Enzymes are very susceptible to thermal inactivation: the higher the temperature, the more rapidly an enzyme is damaged and loses its catalytic properties. The *optimal temperature* for an enzyme-driven reaction is that at which the maximum amount of chemical change is catalyzed. Although an increase in temperature increases the reaction rate, it also shortens the life of the enzyme, so it follows that the optimal temperature must be expressed in relation to the time available for the reaction. Thus, for reactions lasting a few seconds, the optimal temperature may be very high, as thermal inactivation of the enzyme is not important. On the other hand, for reactions lasting much longer, the optimal temperature will be considerably lower since the integrity of the enzyme must be maintained for a greater period of time.

In general, the enzymes involved in metabolic processes in mammals and birds have optimal temperatures in the range 30–40°C at which they are relatively stable, whereas many enzymes found in reptiles, fish, amphibia and invertebrates have optimal temperatures more appropriate to the animals' commonly prevailing temperatures.

The adaptation of the properties of enzymes to make them compatible with an animal's internal thermal environment is probably largely the result of long-term evolutionary processes. However, some remarkable metabolic changes can be instituted within short periods of time during the process of acclimatization.

1.2 Nomenclature

It is common practice to divide the animal kingdom into 'warm-blooded' and 'cold-blooded' species. However, such a subjective segregation of the birds and mammals simply because the surface of their bodies normally feels warm to the touch is grossly unsatisfactory. The terms *homeothermic* (or homoiothermic: Greek homoios=like) and *poikilothermic* (poikilos=various) will be used in this book. Homeothermic forms (birds and mammals) have evolved complex and metabolically expensive means of maintaining the temperature of their body core (see p. 24) within narrow limits. Poikilothermic forms (invertebrates, fish, amphibia and reptiles) have no such mechanisms, thus their temperature generally approximates to that of their environment. Homeothermic animals are sometimes termed *endotherms* since their relatively high metabolic heat production coupled with their low thermal conductance means that their body temperature is largely dependent upon their own oxidative activity. Conversely, poikilothermic animals have a low rate of heat production and a relatively high thermal conductance. Therefore metabolic heat is of less significance than heat from the environment in determining their temperature and they are termed *ectotherms*.

1.3 Animals and their thermal environment

Energy exchange between an animal and its environment is exceedingly complex. At its simplest, it includes on the one hand the exploitation of the chemical energy in the diet and on the other hand the heat exchange with the environment determined by the physical processes of conduction, convection, radiation and evaporation.

Figure 1–3 demonstrates qualitatively the direction of thermal energy exchanges for a homeothermic animal in a moderately warm environment. It is important to remember, however, that in *general* terms very similar considerations will prevail in the case of poikilotherms. This will be discussed in the following chapter.

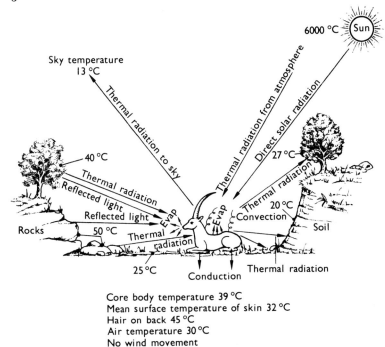

Fig. 1–3 Qualitative representation of the energy exchanges between a homeo-thermic animal and a moderately warm environment. (From GORDON *et' al.*, 1968.)

2 Poikilothermic Animals

Poikilothermic animals remain subservient to their environment, since their activity, and indeed their continued survival, is at all times subject to the prevailing temperature. There are, however, a number of ways by which poikilotherms can exploit the thermal properties of their environment to cause favourable modifications in their body temperature. A discussion of these mechanisms provides the basis of this chapter.

2.1 Aquatic poikilotherms

In many ways, an aquatic environment simplifies the poikilothermic mode of life. Large bodies of water provide a particularly stable thermal environment: diurnal temperature fluctuation is negligible in all but the surface layers and seasonal temperature changes develop slowly. Also, with the exception of water heated by volcanic action, natural bodies of water rarely attain temperatures in excess of 35–40°C. At the other extreme, the concentration of crystalloids in the body fluids of aquatic animals usually renders their tissue freezing point below that of the surrounding water, so that they do not freeze, provided that they remain in the unfrozen water trapped below surface ice (see Chapter 7).

Aquatic poikilotherms therefore rarely face the problem of direct tissue damage from extremes of temperature. Their problem is basically one of ecology, in that their habitat, with its seasonal thermal fluctuations, must be integrated with the requirements of their life cycle.

2.1.1 Aquatic invertebrates

Invertebrates, probably because of their relative structural and biochemical simplicity, can tolerate greater extremes of temperature than can the poikilothermic vertebrates. *Chironomid* (midge) larvae have been reported in hot springs at temperatures of 50°C and, at the other extreme, overwintering insects in sub-arctic regions may survive long periods of sub-zero temperatures (see Chapter 7).

All aquatic invertebrates do not of course survive the extremes quoted above, indeed most individuals can tolerate only relatively small variations from their optimal temperatures. Death due to cold usually occurs long before the tissues freeze and, conversely, excess temperature kills well below the point when protein denaturation threatens.

Experiments investigating the *zone of tolerance* (the temperature spread within which an individual will survive indefinitely) involve keeping animals in tanks at various temperatures and noting their survival times.

Two points of interest arise from such experiments. First, the differences in the tolerance developed by similar animals living naturally in environments with different temperatures. Thus *Terebellids* (polychaete worms) from Greenland die if the water temperature exceeds 6–7°C, whereas the *normal* environmental temperature of such individuals living in the Persian Gulf is 24°C. Secondly, the zone of tolerance of• an individual can be extended in either direction by an acclimatization period in carefully graded temperature baths. The results of one such experiment are shown in Fig. 2–1. Three groups of lobsters were

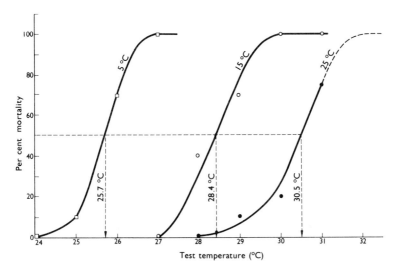

Fig. 2–1   Percentage mortality in three groups of lobsters at various test temperatures after previous acclimatization at 5, 15 or 25°C. (From McLEESE, D. W. (1956). *J. Fish Res. Bd. Can.*, **13**, 247.)

acclimatized to 5, 15 and 25°C in experimental tanks. They were then tested to observe their tolerance to progressively warmer water. The temperature at which 50% of the sample died is indicated in the figure. It can be seen that acclimatization to 25°C meant that half the group could tolerate temperatures of 30.5°C, whereas all of the group acclimatized to 5°C had died by the time that the temperature reached 27°C.

Acclimatization has been shown at the cellular level to be related to a number of changes in factors associated with the metabolism of the animal. These may include increase in O_2 consumption in cold-acclimatized individuals, changes in the rates of specific chemical reactions observed *in vitro* (i.e. in excised tissues), and even changes in the properties of purified enzymes. In summary: acclimatization of *one* individual involves changes in certain fundamental cellular processes accomplished within a short time relative to the animal's life span. The more pronounced variation in temperature tolerance between *different* individuals of closely related species taken from different habitats, although probably embracing in part the changes acquired during acclimatization, reflects in addition more profound genetic variations developed by selective pressure through many generations.

2.1.2   *Aquatic vertebrates – fish*

The temperature of the body of a fish follows very closely that of the water. It can never be less than that of the water because no heat can be lost by evaporation. On the other hand, it is usually impossible for it to exceed for more than a brief period that of the water, because the circulation of blood to the gills, which is extremely efficient in respiratory exchange, unavoidably provides a highly effective means for the thermal equilibration of the blood with the surrounding water.

One notable exception to the generalization that the temperatures of fish do not exceed that of the water is seen in the case of certain large, fast-swimming fish such as the tuna. Figure 2–2 shows the temperature distribution across a tuna measured immediately after capture and recent vigorous swimming. The temperature in the region of the axial swimming muscles may be as much as 12°C above that of the water. This occurs as a result of the large heat production by these muscles when active, coupled

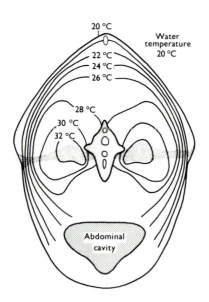

Fig. 2–2   Transverse temperature gradients in a freshly captured 70 kg big-eyed tuna (*Thunnus obesus*). Water temperature 20°C, isotherms plotted at 2°C intervals. (From CAREY, F. G. and TEAL, J. M. (1966). *Proc. Natl. Acad. Sci. U.S.*, 56, 1464.)

with vascular countercurrent heat exchangers (see Chapter 4) which reduce the dissipation of this heat to the environment.

Aquatic invertebrates and fish have a restricted zone of thermal tolerance and characteristic lethal temperatures, and these can be varied within limits by experimental acclimatization or by long-term adaptation of the species to habitats with different thermal limits. One aspect of these mechanisms will be discussed at greater length later in this book: the phenomenon of supercooling in arctic teleosts (Chapter 7).

2.2   Terrestrial poikilotherms

The terrestrial environment provides a severe challenge to the poikilotherm. The annual range of air temperatures in the arctic varies from 20 to 60°C, and *daily* fluctuations between 10 and 45°C are experienced in the desert. In the latter environment the ground temperature in unshaded areas may exceed 80°C. Air has a low specific heat and readily allows the passage of radiant energy. Radiant energy can be absorbed very quickly by the animal from the sun and, on the other hand, it can be lost into space with equal rapidity: a situation which therefore greatly increases the thermal hazard.

Although the body temperature of terrestrial poikilotherms is generally closely related to that of the environment, unlike aquatic forms they can maintain more easily a body temperature differing slightly from that of the environment. Two factors contribute to this: first, the relatively poor thermal conductivity of air, and secondly, the cooling effect of the evaporative loss of body water.

2.2.1   Terrestrial invertebrates

Invertebrates have managed to adapt to and colonize all but the most extreme thermal environments. For reasons of space, however, the discussion of terrestrial invertebrates will be limited to the most versatile and successful group, the insects.

The three main factors that may cause the temperature of an insect's body to vary from that of its environment are outlined below.

1. Absorption of solar radiation   The amount of radiant energy absorbed by an insect depends upon its colour and upon its orientation relative to the sun and thus the area of its surface that it exposes to the direct rays of the sun. Insects may exploit both these factors in regulating solar energy uptake.

The best-documented case of colour changes in relation to temperature is that of the Australian grasshopper (*Kosciuscola*). This insect is a light colour at high environmental temperatures and, as the temperature falls, the insect becomes progressively darker and thus absorbs more of the available solar radiation. The pigment cells responsible are independent effectors directly responsive to temperature: nerves and hormones are not implicated.

Variation of solar energy uptake due to positional changes is well documented in many species. Desert locusts (*Schistocerca*) at low temperatures (less than 17°C) are relatively inactive. Between 17 and 20°C they begin to move and to align themselves in the sun with their bodies at right angles to its rays. At temperatures of between 39 and 43°C they orientate themselves parallel to the sun's rays and at high temperatures raise their bodies off the ground by extending their legs ('stilting'). If the ground temperature becomes excessive they may climb up the vegetation to minimize further heat uptake. 'Stilting' and climbing vegetation are procedures which exploit the rapid decrease in temperature with distance above the hot soil surface. These activities in the locust are examples of *behavioural thermoregulation*.

2. Heat loss by evaporation   Insects are cooled in dry air by evaporation of water from the tracheal system. This may cause a resting insect in a warm dry environment to have a body temperature 3–5°C below that of the environment. It achieves this at the expense of a loss of body water, which may, in adverse circumstances, result in death from desiccation. Normally, negligible amounts of water pass through the cuticle, but at temperatures exceeding 40°C the waterproofing of the cuticle may partially break down, allowing further water loss and hastening death from dehydration.

Heat loss by evaporation does not seem to be capable of modulation by the insect: it is an unavoidable accompaniment of breathing through the spiracles and the degree of evaporation depends on environmental circumstances (see Chapter 4). Although it is tempting to ascribe importance to the increased heat loss by evaporation which would accompany the increased ventilation during flight, the results discussed below indicate that this is of little importance in relation to heat losses by convection.

3. Heat loss during flight   Flight is not immediately possible at low air temperatures because the contraction time of the flight muscles is so slow that sustained flight cannot be achieved. To overcome this difficulty insects indulge in a period of 'warming-up' by fanning the wings before flight commences: the length of this warming-up period depends on the ambient temperature. Thus in the butterfly *Vanessa*, for example, 6 minutes of warming-up was necessary in an 11°C environment, $1\frac{1}{2}$ minutes at 18°C, 18 seconds at 34°C and flight took place immediately at 37°C.

On the other hand, once in flight the insect is faced with the opposite problem: how to dissipate the large amounts of heat produced by the flight muscles.

Heat production during flight can exceed that at rest by a factor of 50, and in insects where the thorax and abdomen have but a narrow connection the thoracic temperature may be 10°C or more higher than that of the abdomen (Fig. 2–3). In the desert locust, prolonged flight

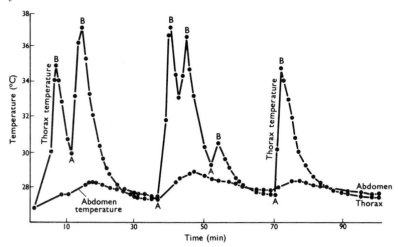

Fig. 2–3 Influence of muscular activity on thoracic and abdominal temperature of a female cecropia moth. A to B indicates periods of wing movements. (From OOSTHUIZEN, M. J. (1939). *J.-ent. Soc. sth. Afr.*, **2**.)

becomes impossible in air temperatures above 38°C because during flight the insect's temperature rapidly approaches the lethal level of 45°C. Flight is intermittent and punctuated by periods of rest for the animal to cool down.

The insect in flight can lose heat in four ways: by conduction, by convection, by radiation or by evaporation of body water (see Chapter 4). It is technically difficult to establish the relative importance of these routes even in insects at rest by virtue of their small size, and the problems of measuring these parameters during flight are even more extreme. However, the use of apparatus such as that shown in Fig. 2–4 has enabled measurements to be taken of the temperature changes within the body during flight under varying conditions of air temperature, wind velocity and relative humidity.

In an elegant series of experiments, CHURCH (1960) estimated the routes of heat dissipation in the locust and other insects. He found that evaporation dissipated only about 5–10% of the heat generated by the wing muscles during flight. Similarly, only 5–15% of the heat was conducted from the thorax to other parts of the insect and about 10% of the heat escaped by radiation. Thus, convection appeared to be the most important means of losing the heat generated during flight and in the locust between 60 and 80% of the heat was lost in this way.

It is of interest to note that the profuse hair growth on the thorax of many nocturnal moths reduces convective heat loss to a considerable degree and thus enables the flight muscle temperature to be maintained in cool night conditions.

Finally, brief mention must be made of the experiments by HEATH and ADAMS (1965) who showed evidence of temperature regulation in the

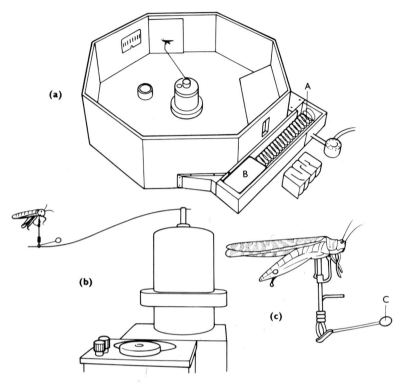

Fig. 2-4 Measurement of temperature changes during flight in the desert locust. (a) General view of flight cabinet with lid removed. Heater (A) and drying unit (B), shown outside the cabinet: relative humidity can be varied between 40 and 95%. (b) Locust mounted on flight mill. (c) Close-up of locust on flight mill arm showing reference thermocouple (C) in plasticene ball. (Diagram (c) from CHURCH, N. S. (1960). *J. exp. Biol.*, **37**, 171.)

sphinx moth during flight. The moth was stimulated to fly, while tethered by the leads from the thermistor recording its thoracic temperature. It was found that the thoracic temperature during flight remained remarkably constant over a wide range of environmental temperatures, while, in contrast, the thoracic temperature in the resting moth was linearly related to the ambient temperature (Fig. 2-5). Even sudden air temperature changes did not produce large changes in thoracic temperature during flight. These observations indicate that the sphinx moth (unlike the locust) is almost homeothermic in flight over a range of environmental temperatures.

It was first thought that the thoracic temperature was regulated by altering the heat production of the flight muscles in response to changes in thoracic temperature, but it has recently been shown that regulation depends upon the conduction of heat to the abdomen. B. Heinrich has

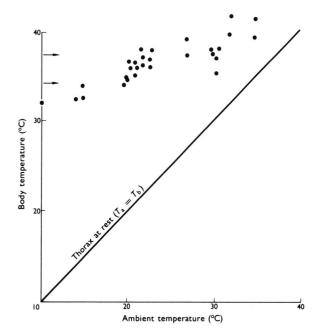

Fig. 2–5 Thoracic temperatures maintained by flying sphinx moth (*Celerio lineata*), compared with the relation between thoracic temperature (T_b) and ambient temperature (T_a) at rest. (From HEATH, J. E. and ADAMS, P. A. (1965). *Nature*, **205**, 309.)

shown that heat produced by the flight muscles is normally carried by the haemolymph to the abdomen where it is dissipated. Tying off the main vessel to the abdomen causes the moths to overheat and to cease flying. However, if the scales covering the thorax are then removed, allowing increased heat loss, the moths will resume flying.

The regulation of heat loss from the thorax would have considerable importance to the maintenance of optimal thoracic temperatures during flight on cool nights while lessening the possibility of overheating in warmer conditions.

SOCIAL INSECTS   Colonial insects, as exemplified by the honey-bee, exhibit elaborate behavioural responses directed towards the maintenance of the optimal temperature of the brood (35°C). If the hive begins to overheat the workers transport water into the hive and aid its evaporation by fanning their wings. This mechanism is effective in maintaining a brood temperature of 36°C in the face of an external temperature of 40°C. In winter, bees cluster together and mutually maintain the temperature of the cluster up to 10°C above that of the environment. In these circumstances the outer bees are much more active than those in the centre of the cluster and there is a continual interchange of bees between the centre and the outer layers.

2.2.2   Terrestrial vertebrates

AMPHIBIA   Amphibia provide perhaps the most extreme example of the importance of evaporation to body temperature. On land, in unsaturated moving air the amphibian behaves as a wet-bulb thermometer. It loses water by evaporation from its moist skin and its body temperature falls below that of its environment.

The potential emergency value of evaporation during heat stress is vividly illustrated by an experiment in which a frog maintained a temperature of 35°C for over 3 h in dry air at 50°C. Such extreme evaporation rates could not of course be maintained for any length of time without access to water and thus amphibia choose habitats which are damp and sheltered so that their *uncontrolled* evaporative loss of water does not constitute a threat to survival.

It must be emphasized that the amphibia cannot regulate evaporation from the skin by physiological means: control is by behavioural mechanisms which ensure a favourable microclimate and thus an acceptable rate of evaporation. Evaporation in the amphibia is in no way comparable with the precisely regulated control of sweating in mammals (see Chapter 4).

REPTILES   Reptiles, with the exception of secondarily aquatic forms such as crocodiles, have accepted the challenge of a completely terrestrial life. They represent a fascinating transition stage with regard to thermoregulation. On the one hand they show many of the behavioural responses described previously for other poikilotherms, while at the same time exhibiting the beginnings of physiological mechanisms which clearly anticipate the thermoregulatory capabilities of the birds and the mammals. Our discussion will be mainly restricted to lizards, since this group has been the subject of most experimental studies.

Lizards kept under laboratory conditions appear to be very un-sophisticated poikilotherms: the body temperature faithfully reflects that of the environment. However, the story is very different if one examines such creatures in their natural habitat. It soon becomes clear that when active, they maintain their temperature within a remarkably narrow range by altering their behaviour to exploit the thermal properties of their environment.

An indication of the '*preferred*' or '*eccritic*' temperature range of any species of lizard can be obtained by measurement of the mean temperature of a number of active individuals in their normal habitat, or by observing individuals in a temperature gradient chamber.

Each species has a characteristic preferred temperature, even though it may exploit habitats with very different climates. Conversely, different species living in the *same* habitat may maintain preferred temperatures which differ by several degrees. It is clear from this that lizards must exert considerable control over their body temperature. This is achieved both by behavioural modifications and by alterations in physiological processes.

1. Behavioural thermoregulation   Most thermal behavioural reactions in lizards ultimately relate to optimizing the uptake of radiant energy from the sun (they are thus sometimes called '*heliotherms*'). Behavioural adaptation can be very complex. For example, the earless lizard found in the south-western U.S.A. spends the cool night period buried in the desert sand. After sunrise, it exposes just its head and waits until the blood flowing through the large sinus has absorbed sufficient heat to raise the temperature of its entire body to the level needed for efficient activity. It then emerges, prewarmed and ready for action. During the day it regulates its uptake of solar energy by periodically sheltering under rocks and by altering its orientation relative to the sun's rays. Many lizards also vary the area of surface exposed to the sun by spreading their ribs and thus altering the shape of the body. Colour changes are also important in some species.

2. Physiological responses   Homeothermic animals respond to changes in deep body temperature by modifying appropriately both metabolic heat production and heat loss from the surface of the body (see Chapter 4). Changes in metabolic rate in most poikilotherms are directly related to body temperature and thus to environmental temperature. Similarly, there is little coordinated control of heat-loss mechanisms. In certain lizards, however, there is evidence both for a change in heat production in a direction appropriate to stabilizing body temperature, and also for regulatory changes in heat-loss mechanisms.

(a) Changes in heat production   Metabolic heat production varies considerably in different reptiles, but in general approximates to that of amphibia and is much lower than that seen in mammals. This endogenous heat production at rest has, as a rule, no significant effect on body temperature. However, certain large lizards such as the monitor lizards (*Varanus*), have been shown to regulate their metabolic heat production in a way which significantly affects body temperature. Indeed, in the Australian monitor (*V. gouldii*), endogenous heat production is the most important factor in determining body temperature and this species can, under appropriate conditions, maintain a metabolic rate which exceeds the basal rate of heat production of a similar-sized mammal. In terms of heat production, therefore, this species bridges the gap between more typical reptiles and the mammals.

A discussion of heat production in reptiles would be incomplete without reference to the female Indian python (*Python molurus*), which can maintain a body temperature as much as 7°C above its environment when incubating its eggs. The eggs are enveloped within coils of the body, spasmodic contractions of which generate the extra heat which keeps the eggs warm (Fig. 2–6).

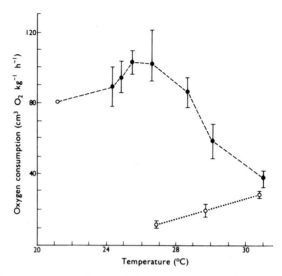

Fig. 2–6   Oxygen consumption at different environmental temperatures in a 14 kg Indian python (*Python molurus*). Upper curve, while brooding eggs; lower curve, during non-brooding period. (From HUTCHISON, V. H. *et al.* (1966). *Science*, 151, 694.)

(*b*) *Heat loss*   Reptiles have poor external insulation: there is little subcutaneous fat and the scales present only slight hindrance to heat transfer. Heat loss can be decreased, however, by slowing the cirulation of blood from the core to the surface tissues. This mechanism is cunningly exploited by the marine iguana (*Amblyrhynchus*) which maintains a body temperature of about 37°C while basking on the rocky Galapagos shore and yet makes periodic excursions into the relatively cold sea to obtain its diet of seaweed. The rate at which the iguana's deep body temperature approaches that of the sea (22–27°C) is greatly decreased by a fall in heart rate. This delay in cooling allows the creature to remain longer in the sea before its reactions are dangerously slowed down, making it an easy prey for sharks. When it emerges from the sea on to the beach, an increase in heart rate facilitates the warming-up process.

Some lizards show the ability to increase heat loss by a panting mechanism. When the body temperature of *Varanus* exceeds 38°C the mouth is held open and the floor of the mouth, and virtually the entire neck region, is involved in deep gulping movements. The increased loss of heat from evaporation of water from the buccal cavity contributes to retarding further rise in body temperature. The thermosensitive part of the reptilian brain, which presumably coordinates such physiological and behavioural responses is, as in animals, the hypothalamus. (A much fuller discussion of the thermoregulatory mechanisms and other aspects of lizard physiology can be found in AVERY, 1979.)

2.3 The evolution of homeothermy

The maintenance of a relatively stable body temperature under a range of environments requires two basic conditions. First, an efficient effector system, that is to say, high metabolic heat production and the means to influence heat loss from the surface of the body. Secondly, heat production and heat loss must be capable of accurate and rapid adjustment in relation to changes in body and environmental temperatures, i.e. there must be a sensitive means of integrating the thermoregulatory effector mechanisms.

The evolution of homeothermy seems to have been limited primarily by the development of the effector side of thermoregulation. Only the birds and mammals have an endogenous heat production of sufficient magnitude to ensure a stable temperature which is well in excess of that of the environment in cold habitats. heat loss in birds and mammals is reduced by effective thermal insulation at the body surface, while efficient vasomotor control of the superficial circulation and the development of sweat glands in mammals allows further regulation of heat loss.

In poikilotherms, heat production is much lower than in homeotherms and the means to regulate heat loss are very poorly developed.

In the absence of an efficient thermoregulatory effector system, sophisticated central integration mechanisms would be inappropriate. Nevertheless, the thermosensitive area of the brain in modern reptiles is probably put to good use in regulating behavioural reactions related to the maintenance of the preferred temperature and also in adjusting such physiological regulatory mechanisms as are present.

It is tempting to speculate that the thermosensitive brain region was developed in the primitive reptiles which were common ancestors to modern reptiles, birds and mammals, and that the evolution of true homeothermy in the latter two groups resulted from the development of efficient effector systems which were accurately integrated by more sophisticated modifications of the primitive reptilian thermosensitive brain area.

3 Energy Balance

3.1 Introduction

The heat produced by an animal is the result of exothermic biochemical reactions in its tissues, the energy for which is derived ultimately from food. Living systems obey the *First Law of Thermodynamics* (the 'Law of Conservation of Energy'), which can therefore be applied to the energy balance of the body. Thus:

TOTAL ENERGY INTAKE = HEAT PRODUCTION + WORK OUTPUT + ENERGY STORAGE (1)

Note: In a mammal in a neutral thermal environment, *total energy intake* means, in effect, *food intake*. However, in certain circumstances, (e.g. a lizard in the hot sun), other forms of energy can contribute to the total energy intake. Note also that some of the chemical energy of the food is not used by the animal and is lost in the urine and faeces (see p. 23).

It is technically impossible to measure all the four variables accurately, so the equation is simplified by measuring heat production in a resting subject who has been starved for at least 18 h. Thus:

total energy intake = 0 (the last meal will have been digested
within 18 h)
work output = 0 (the subject is resting)

and the equation becomes:

HEAT PRODUCTION = − ENERGY STORAGE (the subject is burning
up his energy stores). (2)

The measurement of heat production under these conditions indicates the *basal metabolic rate* (B.M.R.). The B.M.R. gives a measure of the heat production of the body when it is 'ticking-over': it is a valuable clinical tool and it provides a base-line from which factors influencing heat production can be assessed.

3.2 Calorimetry

DIRECT CALORIMETRY It is technically very difficult to measure accurately the heat production of an experimental subject. Such measurements are made in special, well-insulated chambers called *calorimeters* (Fig. 3–1). Water flows through coils of copper pipes within the calorimeter and absorbs the heat produced by the subject. If the temperature increase of the water is measured, together with the rate of flow, the heat production in kilojoules (kJ) or kilocalories (kcal) can be calculated. To this must be added the latent heat present in the water

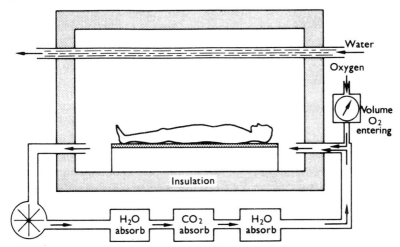

Fig. 3–1 The human calorimeter. In a resting subject the total energy output is the sum of (i) the heat evolved (measured from the temperature rise of the water flowing in coils through the chamber) and (ii) the latent heat of vaporization (measured from the amount of water vapour extracted from the circulating air by the first H_2O absorber). CO_2 must be absorbed to prevent its accumulation within the chamber; this process evolves water, so a second H_2O absorber is needed. Oxygen consumption can be measured by noting the rate at which O_2 must be added to keep the chamber in a steady state. (From BROWN, A. C. and BRENGELMANN, G. Temperature regulation and energy metabolism. In RUCH and PATTON, 1973.)

vapour of the perspiration and expired air. The water vapour produced is measured from its absorption in sulphuric acid: each gram of water accounts for 2.451 kJ (0.585 kcal) (the latent heat of vaporization of water at 20°C).

INDIRECT CALORIMETRY   Calorimeters, although simple in principle, are very expensive and complex in operation and there are very few in existence. Fortunately, however, it is possible to calculate heat production *indirectly* from simple measurements of O_2 consumption, CO_2 production and nitrogen excreted into the urine.

The expression $\dfrac{CO_2 \text{ production}}{O_2 \text{ consumption}}$ is known as the *respiratory quotient* (R.Q.) and it can be related to the composition of the diet and thus to the proportions of carbohydrate, fat and protein being oxidized.

When carbohydrate is oxidized, the consumption of O_2 by volume is exactly equal to the CO_2 production, thus the R.Q. $= 1$. For example: GLUCOSE:

$$C_6H_{12}O_6 + 6O_2 \rightarrow 6CO_2 + 6H_2O$$

1 gram molecule of glucose (180 g) requires 6×22.4 l (134.4 l) of O_2 and produces 26×22.4 l (134.4 l) of CO_2 and 2820 kJ (673 kcal) of heat.

Thus the R.Q. $= \dfrac{134.4}{134.4} = 1$ and the *calorific value* of 1 l of O_2 when glucose

is oxidized is $\dfrac{2820}{134.4}$ kJ $= 20.9$ kJ $\left(\dfrac{673}{134.4} \text{ kcal} = 5.0 \text{ kcal} \right)$.

When fat is oxidized the volume of O_2 used exceeds the volume of CO_2 produced. For example: TRIPALMITIN:

$$C_{51}H_{98}O_6 + 72.5 O_2 \rightarrow 51 CO_2 + 49 H_2O$$

1 gram molecule of tripalmitin (806 g) requires 72.5×22.4 l (1624 l) of O_2 and produces 51×22.4 l (1142 l) of CO_2 and 32 083 kJ (7657 kcal).

Thus the R.Q. $= \dfrac{1142}{1624} = 0.703$ and the calorific value of 1 l of O_2 is

$\dfrac{32\,083}{1624}$ kJ $= 19.7$ kJ $\left(\dfrac{7657}{1624} \text{ kcal} = 4.7 \text{ kcal} \right)$.

It is more difficult to calculate the R.Q. for protein, because some of the oxygen and carbon of the constituent amino acids remain combined with nitrogen and are excreted as nitrogenous compounds in the urine and faeces. However, it has been found that a reasonably accurate estimate of protein metabolized can be derived from knowledge of the nitrogen excreted in the urine. Thus 1 g of urinary nitrogen indicates the catabolism of 6.25 g of protein. It has also been shown that 1 g urinary nitrogen signifies that:

5.94 l O_2 have been consumed $\Big\}$ (R.Q. 0.081)   (3)
4.76 l CO_2 have been produced $\Big\}$ (4)
111.1 kJ (26.51 kcal) have been produced.   (5)

Armed with these facts, it is possible to calculate the heat production of a subject by making three relatively simple measurements.

(a) CO_2 produced $\Big\}$
(b) O_2 consumed $\Big\}$ during the experimental period.
(c) N in the urine $\Big\}$

This method of *indirect calorimetry* will be explained by reference to the following example:

$$\text{urinary N/h} \qquad\qquad = \;\; 0.37 \text{ g}$$
$$\text{total } CO_2 \text{ produced/h} = \;\; 9.48 \text{ l}$$
$$\text{total } O_2 \text{ consumed/h} = \;\; 11.40 \text{ l.}$$

Proportion of the total O_2 used in protein oxidation (see equation 3 above
$$= 0.37 \times 5.94 = 2.20 \text{ l.}$$
Proportion of the total CO_2 resulting from protein oxidation (see 4 above)
$$= 0.37 \times 4.76 = 1.76 \text{ l.}$$
Heat produced by oxidation of protein (see equation 5 above)
$$= 0.37 \times 111.10 = 41.10 \text{ kJ (9.81 kcal).}$$

Subtraction of the CO_2 produced and O_2 consumed in protein oxidation from the total CO_2 production and O_2 consumption reveals the CO_2 produced and O_2 consumed in oxidizing fat and carbohydrate. Thus:

CO_2 produced from fat and carbohydrate $= 9.48 - 1.76 = 7.72$ l
O_2 consumed by fat and carbohydrate $= 11.40 - 2.20 = 9.20$ l.

The *non-protein respiratory quotient* (N.P.R.Q.) is thus $\dfrac{7.72}{9.20} = 0.84$.

Knowing the N.P.R.Q. it is simple to read off from Fig. 3–2 the calorific value of each litre of oxygen used in the oxidation of the fat and carbohydrate by the subject.
Calorific value of 1 l O_2 when R.Q. $= 0.84$ is 20.28 kJ (4.84 kcal), see Fig. 3–2. Thus:

heat produced from the oxidation of fat and carbohydrate
$= 20.28 \times 9.20 = 186.56$ kJ (44.53 kcal)

heat from protein oxidation $= 41.10$ kJ (9.81 kcal), (see above)

total heat production/h $= 227.68$ kJ (54.33 kcal).

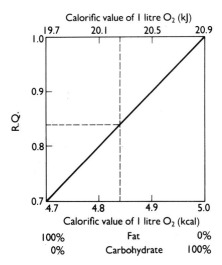

Fig. 3–2   Relation between respiratory quotient and the calorific value of oxygen. (At an R.Q. of 0.84 the calorific value of 1 l of O_2 is 4.84.)

Estimates of heat production by indirect calorimetry agree well with those obtained by the more laborious means of direct calorimetry.

3.3   Basal metabolic rate

Once the B.M.R. in kJ or kcal/24 h has been estimated by either direct or indirect calorimetry, the size of the individual must be taken into account. Table 1 shows a comparison of B.M.R. in kcal/24 h expressed in terms of body weight, surface area and (body weight, W)$^{0.73}$ for four different species. It can be seen that a comparison in terms of body weight is unsatisfactory and yields higher values for smaller subjects. Comparison in terms of surface area gives some improvement, but it is difficult to estimate accurately the surface area of an individual. It has been found, however, that the heat production, divided by the weight raised to the power of 0.73, yields an answer of about 293 kJ/24 h or 70 kcal/24 h, irrespective of the size and species of the subject.

Table 1   Comparison of B.M.R. in four different species: kJ/24 h (kcal/24 h).

	Body wt (kg)	B.M.R./kg	B.M.R./M^2	B.M.R./W$^{0.73}$
Pig	128.0	90.0 (19.1)	4517 (1078)	297.6 (70.8)
Man	64.3	134.5 (32.1)	4366 (1042)	295.6 (70.5)
Dog	15.2	216.7 (51.5)	4353 (1039)	297.1 (70.9)
Mouse	0.018	2740.0 (654.0)	4978 (1188)	300.0 (71.6)

CONDITIONS WHICH INFLUENCE B.M.R.   B.M.R. in man depends on the age and sex of the individual; it is highest in infants and declines progressively with age and at any age tends to be slightly higher in males than females. Under-nutrition or starvation and underactivity of the thyroid gland lower the B.M.R., whereas overactivity of the thyroid gland and fever tend to elevate the B.M.R.

The calorie intake in the diet necessary to support basal metabolism is about 2000 kcal/24 h for an average-sized male subject (i.e. approximately 8400 kJ/24 h). From equation 1 on p. 18 it follows that if the dietary intake of calories is less than this the subject loses weight; conversely, if the dietary intake exceeds the energy requirement, weight is gained.

FACTORS WHICH RAISE THE HEAT PRODUCTION ABOVE THE BASAL LEVEL
(a) *Muscular effort*   Physical exercise requires fuel so that, as is well known, people employed as labourers or lumberjacks need to eat more than people in sedentary occupations. A lumberjack, for example, requires in his daily diet 13 000 kJ (3000 kcal) in addition to the 8400 kJ (2000 kcal) needed for his basal metabolism, while a sedentary student requires only an extra 2100 kJ (500 kcal).

(b) *Mental effort*   Thought processes require negligible dietary energy. As a result of experiments where subjects were required to perform difficult mental arithmetic, it has been calculated that the extra energy required for this would be met by half a salted peanut per hour!

(c) Feeding   The ingestion of food increases heat production. This increase is known as the *specific dynamic action* (S.D.A.) or the calorigenic action of food. Heat production begins to rise about 1 h after feeding, reaches a peak after about 3 h and then remains above the basal level for several more hours. Protein has a high S.D.A.: 30% of the calorific value is lost as heat. Therefore, if an animal required, for example, 100 kJ/day in its diet to maintain its energy balance, it would have to be given 130 kJ/day if its diet were pure protein. Corresponding figures for carbohydrate and fat are 106 and 104 kJ/day.

The mechanism of the S.D.A. is not known, but it is clear that it does not derive from events within the digestive tract, because amino acids infused intravenously produce the same S.D.A. as when they are fed orally. The S.D.A. must therefore be a consequence of metabolic events after the food products have been absorbed into the circulation and distributed to peripheral tissues. The heat produced contributes to the maintenance of body temperature.

3.4   The efficiency of feeding

During digestion not all the energy of the diet is made available for the production of work or the storage of energy by the body. The efficiency of utilization of the diet can be computed as shown below.

Gross energy = heat of combustion of the food as measured in a bomb calorimeter.

Digestible energy = gross energy − heat of combustion of faeces.

Digestible energy gives a measure of the calorific value of the food absorbed from the digestive tract. However, not all this food is completely oxidized in the body: protein forms urea and other nitrogenous compounds excreted in the urine.

Thus the energy available for metabolism is:

available (metabolizable) *energy* = digestible energy − heat of combustion of the urine.

In man, on a typical mixed diet, the *available energy* is about 85% of the *gross energy*.

A proportion of the available energy is expended as the S.D.A., so the *net energy* (that which can contribute to work or energy storage) is less than the available energy. In practice, about 20% of the available energy is lost as S.D.A. so that the overall efficiency of feeding $\dfrac{\text{net energy}}{\text{gross energy}} \times 100$ is about 68%.

There is considerable variation between human subjects in the efficiency of feeding which can in part be explained by genetic differences.

4 Heat Production and Heat Loss

4.1 Introduction

In Chapter 2 it became apparent that the body temperature of poikilotherms could vary considerably, dependent upon environmental conditions. Homeotherms, on the other hand, contrive to maintain a remarkably constant body temperature throughout a wide range of environmental temperatures (Fig. 4–1).

It must be emphasized, however, that the term 'body temperature' in this context means the temperature of the *core* of the body, that is to say, tissues lying deeper than about 2.5 cm from the surface. The temperature of more superficial parts of the body is much more variable, and under extreme conditions may differ from the core temperature by many degrees (Fig. 7–2).

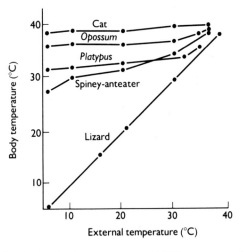

Fig. 4–1 Variation in the body temperature of different types of animals kept for 2 h in environmental temperatures between 5 and 35°C. (From MARTIN, C. J. (1930). *Lancet* (2), **108**, 561.)

The *core temperature* (conveniently measured with a rectal thermometer) varies in different species and tends to be higher in birds than mammals; 44°C in the great titmouse, 42.5°C in the fowl, 39°C in the pig, 38°C in the cow and 37°C in man. Marsupial and monotreme mammals tend to have lower and more variable core temperatures. There are diurnal variations in these temperatures: thus in man the temperature at 4 a.m. may be as much as 1°C lower than at 6 p.m. and, of course, core temperature may be temporarily elevated in disease or during extreme exercise.

The regulation of core temperature in homeothermic animals rests upon a very simple principle: the temperature will remain constant if

HEAT PRODUCTION = HEAT LOSS

This chapter will be concerned with the mechanisms for heat production and heat loss, while Chapter 5 will consider the way in which these mechanisms are integrated to ensure a stable core temperature.

4.2   Heat production

Heat production mechanisms can conveniently be divided into two categories: those which involve skeletal (voluntary) muscles and those which do not. The latter mechanisms are sometimes termed '*non-shivering thermogenesis*'.

4.2.1   Heat production by skeletal muscle

One of the most obvious responses to a cold environment in both birds and mammals is *shivering*. Shivering is an involuntary tremor of skeletal muscle produced by impulses passing down somatic motor nerves to the muscle. The autonomic nervous system is not implicated. Shivering can increase heat production to between two and five times the basal level and thus provides a quickly variable source of additional heat production of particular value to temperature regulation.

Skeletal muscle is also implicated in the general increase in voluntary movement brought about by cold discomfort in such manoeuvres as stamping the feet, flapping the arms and so on.

Exercise, of course, results in a much larger increase in heat production (up to 20 times the basal rate during vigorous exercise), but this extra heat is rarely called upon during thermoregulation. In fact, under most circumstances it actually impairs thermoregulation and necessitates taking steps to increase the loss of heat from the body to preclude dangerous overheating.

4.2.2   Non-shivering thermogenesis

Two important components of non-shivering thermogenesis, already considered, are the B.M.R. and the S.D.A. Thus, there is a steady production of heat even in a resting fasting subject which is supplemented by additional heat production after meals. Neither of these processes are materially influenced by the heat production of skeletal muscle.

CALORIGENIC ACTION OF HORMONES (*a*) *Thyroid hormones* It was noted previously that underactivity of the thyroid gland was associated with a low B.M.R., while the B.M.R. was high when the thyroid was overactive. *Thyroxine* and *triiodothyronine*, the thyroid hormones, increase the oxygen consumption and hence the heat output of almost all metabolically active tissues. The effect of an injection of thyroxine begins after a latent period of several hours, but may last for six or more days.

Cold is a potent stimulus to thyroid function: it activates areas of the hypothalamus (see Chapter 5) causing the release of the polypeptide, *thyrotropin releasing factor* (T.R.F.). T.R.F. passes down the portal vessels linking the hypothalamus and the anterior pituitary gland and stimulates the release of the *thyroid stimulating hormone* (T.S.H.). T.S.H. passes into the blood and is transported to the thyroid gland where it promotes the synthesis and release of thyroxine (T_4) and triiodothyronine (T_3), which elicit an increase in metabolic rate, which, in turn, assists the animal to respond to the initial cold stimulus. The prolonged action of thyroid hormones on metabolism makes them particularly useful in the response of an animal to long-term exposure to cold.

(b) *Adrenal medullary hormones*   An immediate and relatively short-lived increase in heat production follows the injection of adrenaline. This is due to an increase in cellular oxidations in tissues in general and also to an increased utilization of carbohydrate consequent upon the increase in blood glucose concentration evoked by adrenaline. It is probable that the immediate increase in non-shivering thermogenesis following exposure to cold can be largely attributed to the release of adrenaline since, when exposed to cold, animals deprived of their adrenal glands begin to shiver much earlier than control animals.

BROWN FAT   New-born animals from certain mammalian orders, including primates and rodents, have a unique and versatile source of heat production in the form of pads of brown fat. Brown or multilocular fat differs histologically and functionally from the usual white or unilocular fat. In brown fat cells the fat droplets are numerous and are interspersed with many mitochondria. There is a rich blood supply and a generous sympathetic nervous innervation.

Brown fat has been likened to an electric blanket and this is quite a justifiable analogy, for in a cold environment the heat production from the brown fat is switched on by impulses passing down the sympathetic nerves. It is also stimulated by noradrenaline release from the adrenal medulla. When the cells are active, the free fatty acid stores are burned up within the cells with the aid of mitochondrial enzymes and heat is produced. Furthermore, the proximity of major thoracic blood vessels ensures that most of the heat produced is carried into the core of the body.

Brown fat deposits are well marked in hibernators just prior to hibernation and indeed for many years brown fat was known as the 'hibernating gland'. It seems to be of particular importance during arousal from hibernation (see Chapter 7).

The pathways involved in the control of heat production are summarized in Fig. 4–2.

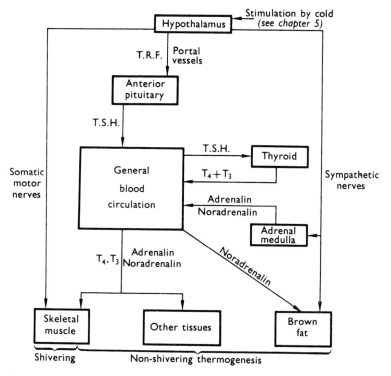

Fig. 4-2   Diagram to show the main pathways by which heat production is regulated. T.R.F.=thyrotropin releasing factor, T.S.H.=thyroid stimulating hormone (thyrotropin), T_4=thyroxine, T_3=triiodothyronine.

4.3   Heat loss

Four physical processes are implicated in the transfer of heat from the core of the body to the environment: conduction, convection, radiation and evaporation. Ignoring the 5% lost in the urine and faeces, heat loss takes place through conduction, convection, radiation or evaporation from the skin surface and by evaporation from the respiratory tract (see Fig. 1-3).

4.3.1   Heat loss from the skin

Homeotherms produce heat continually at such a rate that the temperature of the core of the body is normally above that of the ambient air. Heat is dissipated to the environment from the skin surface which therefore is cooled, establishing a gradient of temperature from the body core to the skin surface. However, this temperature gradient between body core and skin surface will be changed by alteration in the thermal conductivity of the subcutaneous layers, brought about by changes in the blood flow through these regions.

BLOOD FLOW IN THE SKIN   Under conditions of decreased core temperature (*hypothermia*), when it is essential to minimize heat loss, blood flow to the superficial layers of the body is severely restricted and the thermal conductivity of the outer 2–3 cm or so of the body becomes comparable with that of cork. Transfer of heat from core to surface is thus reduced and the skin surface temperature falls, causing a reduction in heat loss by convection and radiation (see below).

When the core temperature is high (*hyperthermia*), and it becomes necessary to increase heat loss through the skin, there is a massive increase in blood flow through the superficial layers of the body. This may be up to 100 times the minimum flow. Under these circumstances the thermal distinction between the core of the body and the surface is virtually lost as the skin temperature approaches that of the core. Heat transfer from the surface is facilitated as the higher skin surface temperature aids loss of heat by convection and radiation.

It is clear from the preceding remarks that alterations in the blood flow to the skin are of fundamental importance to the regulation of heat loss. The control of the blood vessels in the skin will therefore be considered in some detail.

Figure 4–3 illustrates diagrammatically the differences in the patterns of blood flow through an extremity under conditions of heat conservation and of heat dissipation.

Heat loss from the skin is minimized by restricting the blood flow through superficial vessels to the minimum necessary for the metabolic needs of the tissues (Fig. 4–3a). Most of the blood passes through the relatively high resistance capillary bed and little is short circuited through the arterio-venous anastomoses (low resistance channels between the arterial and venous sides of the circulation: these are relatively thick-walled and little metabolic exchange takes place). Venous blood is returned predominantly by the *venae comites*, which are deep venous pathways running in close proximity to the arterial vessels. This anatomical arrangement allows *countercurrent heat exchange* to take place. Arterial blood at high temperature flows from the core towards the periphery, while venous blood, which was cooled at the periphery, flows in the opposite direction. Heat passes from the arterial blood across to the adjacent venous blood, thus the arterial blood is precooled by the time it reaches the capillary beds, whereas the venous blood has been substantially rewarmed by the time it re-enters the core. Efficient countercurrent heat exchange is of great value to the conservation of heat, since a large proportion of the heat carried by arterial blood is short circuited back into the core without ever reaching the periphery. The efficiency of the vascular countercurrent heat exchanges depends upon the relative anatomic configuration of the arterial and venous elements.

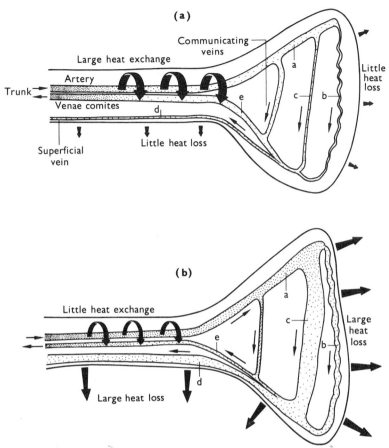

Fig. 4–3 Blood flow through an appendage when directed towards (a) heat conservation, and (b) heat loss. In (a): a, absolute flow through superficial tissues low, thermal conductivity equivalent to cork; b, flow within skin restricted to capillaries; c, little flow in arterio-venous anastomoses (shunt pathways); d, little flow in superficial veins; e, cold blood returns from the extremities in the venae comites; it is warmed by counter-current heat exchange from the arterial blood. In (b): a, absolute flow through superficial tissues high; b, some increase in capillary flow; c, large blood flow through arterio-venous anastomoses; d, most blood returns through superficial veins; e, little blood returns through venae comites. Intensity of shading indicates temperature of blood, width of vessels indicates relative blood flow.

In the flippers of whales and porpoises the arteries are completely surrounded by venous channels, allowing the maximum surface area for heat exchange while many other species have highly complex intermingling of arterial and venous elements at the roots of the appendages: these areas are known as the *retia mirabilia* (miraculous networks). Countercurrent heat exchange is particularly significant in animals living in conditions of extreme cold, thus penguins can remain with their feet in contact with ice for long periods without a dangerous loss of heat. In man there are no distinctive anatomical concessions to vascular heat exchange, but the general tendency for venous elements to run close to the corresponding arteries favours heat exchange.

When it becomes necessary to promote heat loss there is an increase in the absolute blood flow through the skin (Fig. 4–3b). Most of the blood is channelled through the low resistance arterio-venous anastomoses and serves to increase the surface temperature as well as the flow of heat from core to surface. Blood returns to the core predominantly by means of the superficial veins, which allows further opportunity for heat to pass to the environment.

CONTROL OF SKIN BLOOD FLOW   The changes in the pattern of blood flow through the superficial layers are largely controlled by activity in nerves from the sympathetic division of the autonomic nervous system. There are three mechanisms which may be implicated to various degrees in the widening of vessels (vasodilation) and consequent increase in blood flow in response to hypothermia. First, a decrease in the activity of nerves which cause cutaneous blood vessels to constrict in the cold (vasoconstrictor fibres): secondly, an increase in the activity of sympathetic vasodilator fibres (these release acetyl choline and have a restricted distribution); thirdly, the vasodilator action of *bradykinin*, a substance formed from tissue proteins by an enzyme produced by active sweat glands (sympathetic *sudomotor* nerves activate sweat glands: see p. 32). The way in which the activity in these three sympathetic pathways is regulated by hypothalamic centres will be discussed in the following chapter.

EXTERNAL INSULATION   In addition to being the major pathway for controlling skin blood flow, the sympathetic nervous system plays a further role in the regulation of heat loss, since it determines the effectiveness of the external insulation provided by fur or feathers. Air is a very poor conductor of heat and in consequence a layer of stagnant air trapped at the surface of the body comprises a very effective barrier to heat loss. Both fur and feathers provide a means of trapping air at the body surface, and the effectiveness of this insulating coat can be controlled to some degree by altering the angle of the hairs or feathers relative to the skin and thus varying the thickness of the layer of trapped air.

In mammals, individual hairs are pulled into a more upright position (piloerection) by the contraction of slips of smooth muscle brought about by impulses in the sympathetic nerves supplying them. In man, where the investment of body hair has been effectively lost, contraction of the piloerector muscles is signalled by the appearance of 'goose-flesh' (horripilation). Man, however, has compensated for the absence of an insulating fur coat by developing artificial insulation in the form of clothes and by modifying his environment by constructing shelters and using fires. The relative effectiveness of clothing in man and fur in other species will be considered in Chapter 7.

4.3.2   Physical processes in heat loss

CONDUCTION AND CONVECTION   The rate of loss of heat by conduction depends upon the temperature difference between the skin and the environment and upon the area of contact. It is normally the least important of the four paths of heat loss, because of the low thermal conductivity of air and the limited area of contact between the body and the solid environment. Under special circumstances, such as contact with a cool object, cold ground or immersion in water, conductive heat loss becomes more significant.

Much more important is convection, because the low specific heat of air allows air in contact with the skin to be warmed rapidly. The warm air is less dense and rises and is replaced by cooler air, thus setting up convection currents and maintaining the temperature gradient between the skin and the adjacent air. The effectiveness of convection depends upon the temperature gradient between the skin and the adjacent air and upon the free movement of air at the skin surface. The latter process is facilitated by air movements so that one feels cold in a draught and is cooled more rapidly on a windy day than on a still day, even if the air temperature is identical on both days. However, part of this effect can be attributed to increased evaporation of sweat (see below). Heat loss increases with the square root of the wind speed up to speeds of about 110 km/h; little further increase in heat loss is seen at higher wind velocities.

Heat loss by convection is diminished by factors which impede air movement at the skin surface. Fur, feathers and clothing serve this purpose, while postural changes can minimize exposure to air movement, as do behavioural responses such as seeking or constructing shelter.

RADIATION   Radiant heat is transmitted by electromagnetic waves, which are converted to heat when they fall upon cooler objects which absorb them. Similarly, a hot object can lose heat by emitting radiant energy. In a comfortable temperate environment loss by radiation comprises about 50% of the total heat loss in man. The human body, irrespective of colour, radiates heat in the near infrared between wavelengths of 90 000 nm and 200 000 nm. Air is a poor absorber of heat, so the body does not radiate heat to the air, but rather to solid objects in the environment. For this reason, the parts of the body absorbing radiant heat from a hot object, such as the sun or an electric fire, are warmed, while those parts radiating heat to a colder object, such as a window, are cooled. The amount of heat radiated depends upon the difference in temperature between the skin and the solid environment and upon the area of skin exposed. Clothing or fur radiate less heat than skin, since such covering is generally at a lower temperature than the skin.

The effective radiating surface of the naked human body comprises about 85% of the total surface area, since apposed surfaces, such as the inner surfaces of the upper arms and the lateral thorax and the inner surfaces of the thighs, radiate to each other and not to the environment. It follows also that huddling-up in the cold and spreading one's limbs in the heat modify radiant emission.

The primary physiological mechanism which determines radiant heat loss is the regulation of blood flow through the skin and hence of the skin temperature.

EVAPORATION   Each gram of water evaporated from the surface of the skin at room temperature entails the loss of about 2.45 kJ (0.58 kcal) from the body. Evaporative heat loss can be divided into two categories: *insensible water loss* from the skin and respiratory tract and *regulated loss* by thermal sweating and panting.

1. Insensible water loss   Under normal or basal conditions man produces no sweat, with the possible exception of slight secretion from the *apocrine glands* (see below). However, about 30 g of water are vaporized each hour, representing a loss of 30×2.45 kJ. This is called insensible water loss and is the result of diffusion of water through the skin in the vapour phase (it does not wet the skin) and the loss of water in the expired air from the respiratory tract.

2. Regulated evaporative heat loss   *(a) Thermal sweating*   There are two types of sweat gland in man. The first, the *apocrine glands*, are restricted to the axillary and pubic regions and the hands and feet; they play no significant role in thermal sweating. The second, the *eccrine glands*, have a wide distribution over the body and number about 2 500 000. They are innervated by sympathetic nerve fibres and secrete only when stimulated. Eccrine sweat is a dilute salt solution, formed by the ultrafiltration of blood plasma in the glands and, although some sodium and chloride ions are subsequently reabsorbed, the secretion of sweat can represent a considerable loss of salt as well as water from the body.

Under extreme conditions, sweating rates of over 4 l/h have been recorded for short periods, and, in experiments by the U.S. army in Death Valley, California, subjects readily produced up to 12 l of sweat per day. In these experiments there was, of course, adequate salt and water intake.

Sweat is of little value in cooling the body unless it is evaporated, so that environmental factors such as the relative humidity and the speed of air movement play a major role in determining the efficiency of sweating. Sweating begins in man at an environmental temperature of about 29°C. Ideally one should be unaware of sweating since the accumulation of sweat on the skin is the accumulation of water which is not removing heat and is associated with the onset of discomfort: this is common in humid conditions.

(b) *Panting*   Sweating clearly provides a means of cooling the body
even when the environmental temperature exceeds that of the body. What
mechanisms are available then in animals in which sweat glands are
scarce or absent? Birds have no sweat glands, while in many mammalian
species with heavy coats of hair such as cats and dogs, sweat glands are
only found on the pads of the feet. Evaporative heat loss in these animals
is brought about by increased air movement over the moist mucosal
surfaces of the mouth and upper respiratory tract. There is rapid shallow
breathing, sometimes called *polypnoeic panting*, associated with increased
salivation. Such a pattern of breathing establishes efficient air flow across
the moist surfaces without causing a dangerous decline in the partial
pressure of CO_2 in the blood, such as would result from rapid *deep*
breathing. The respiration rate reaches very high levels in extreme
panting: 200–300 respirations per minute in the dog and more than 600
per minute in some birds (the resting respiration rate in man is about 14
per minute). A further means of bringing about evaporative heat loss is
demonstrated by the kangaroo. At high environmental temperatures a
copious flow of saliva ensues and the animal distributes this over the
surface of the body by licking its legs, feet, tail and ventral surface. In
other species, such as the hippopotamus and pig, wallowing aids heat
loss, both by direct conduction to the cool mud or water, and by the
subsequent evaporation when the animal emerges.

4.3.3   Conclusion

Heat loss takes place by conduction, convection, radiation and
evaporation. With the exception of evaporation, the rate of heat transfer
along these pathways is proportional to the temperature gradient, i.e. the
difference between skin and environmental temperature. It follows,
therefore, that as the environmental temperature rises, the heat loss by
conduction, convection and radiation will decrease, until no heat is lost
by these routes and indeed the heat flow may be reversed. Under these
conditions evaporation provides the *only* means of losing heat.

The sympathetic nervous system is of fundamental importance as a
pathway by which heat loss can be controlled. It is the major influence on
skin blood flow and thus on skin temperature, and it also controls
thermal sweating.

5 Thermoregulatory Control Systems

5.1 Introduction

We have seen that the maintenance of a stable core temperature requires that the heat production of the body should equal the heat loss from the body. In order to achieve such a balance it is necessary to have a control centre which can integrate incoming sensory information about core temperature and then modify appropriately the heat production and heat loss mechanisms. However, this is only part of the story, for the thermal inertia of the body is such that sudden changes in environmental temperature would result in considerable loss or gain of heat before appreciable changes in core temperature occurred and thus control mechanisms were set working. There is, therefore, another line of defence, provided by temperature receptors on the body surface. These peripheral receptors send information to the control centres, which allows *anticipatory* changes in heat production or heat loss to take place in advance of changes instituted by the control centres in response to alterations in core temperature.

In homeotherms, as mentioned briefly in Chapter 4, such thermoregulatory control centres are situated in the hypothalamus. It will be the purpose of this chapter to consider in more detail the functioning of the hypothalamic centres and to describe some of the experiments which have helped to clarify their mode of action.

5.2 Temperature receptors

It is apparent from one's own experience that there are *peripheral thermoreceptors* situated in all parts of the skin and in the mouth and the upper part of the alimentary canal. These skin receptors provide us with most of our conscious sensations of temperature, but, in addition, information from these peripheral receptors has considerable influence on the hypothalamic control centres.

In addition to peripheral thermoreceptors, there is now conclusive evidence for the existence of temperature-sensitive elements in the body core itself. One simple experiment will suffice to indicate the existence of such *central thermoreceptors*. If one arm of a human subject is immersed in hot water while the other arm is left exposed to air, the exposed arm soon begins to sweat and to turn pink as the blood flow to the skin increases (vasodilation). This response in the exposed arm might be the consequence of nervous information from the peripheral thermo-receptors in the skin of the arm in hot water, or it might be due to the warming of blood in the heated arm which in turn might activate central thermoreceptors. This question can be resolved by the simple experiment of occluding the circulation to the heated arm with an inflatable cuff placed round the upper arm. When the flow of warmed blood from the heated arm is obstructed, the sweating and vasodilation in the other arm subsides and only resumes when the cuff is released. (*Note:* You should not attempt this experiment yourselves.) This experiment shows that the peripheral thermoreceptors in the heated arm are not responsible for the responses in the exposed arm since nerve impulses from the thermoreceptors would not be blocked by the inflated cuff. The response in the exposed arm must therefore be due to the heated blood from the hot arm activating some central thermoreceptors in the body core. Thus the response ceases when the cuff is inflated and heated blood cannot return to the trunk.

Such an experiment gives little indication as to the whereabouts of the central thermoreceptors, but we shall see from section 5.4 that the results of direct thermal stimulation have now established that they are in the hypothalamus.

5.3   Thermoreceptor interactions

The control centres receive information from both central and peripheral thermoreceptors, so it is important to establish how this information is integrated to produce appropriate thermoregulatory responses.

Until comparatively recently it was thought that a rise in skin temperature was necessary to induce sweating. Such a view is a little surprising when it is remembered that sweating tends to reduce skin temperature, for this means that the system would be self-limiting. It is now clear that sweating is primarily a response to an increase in core temperature. This has been established from some ingenious experiments on human subjects where the rate of sweating was measured while the skin and core temperatures were driven in opposite directions (Fig. 5–1).

It can be seen that sweating followed the core temperature, so that when the subject lowered his core temperature by swallowing ice, sweating was reduced, despite the fact that the already high skin temperature rose even further at this time as a consequence of the reduced sweating.

From this experiment one might perhaps question the usefulness of peripheral thermoreceptors to the control of sweating, but it is now known that skin receptors do influence the onset of sweating at low environmental temperatures. Figure 5–2a shows the relation between cranial temperature and sweating rate in man at different environmental temperatures. At environmental temperatures above 33°C sweating begins as soon as the cranial temperature reaches 36.8°C and increases proportionately as the cranial temperature rises. However, in a colder

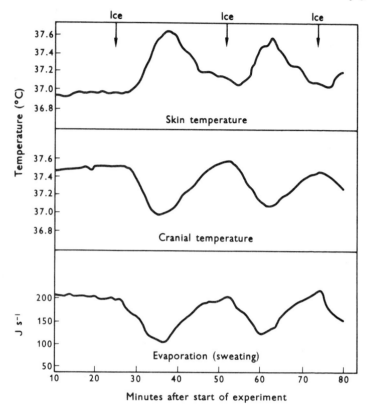

Fig. 5–1 Relation between sweating and brain temperature in a warm human subject. Internal temperature, measured at the eardrum provides an acceptable index of hypothalamic temperature. From 0–25 min subject equilibrates in warm chamber (45°C). Ice ingested at 25, 52 and 75 min cools the blood and hence lowers eardrum (brain) temperature. Note parallel decline in evaporation and total heat loss. Skin temperature rises as sweating is inhibited. (Modified from BENZINGER, T. H. (1959). *Proc. natn. Acad. Sci. U.S.A.*, **45**, 645–59.)

environment (29°C), the onset of sweating is delayed until the cranial temperature reaches 37.5°C. Thus it appears that the onset of sweating can be inhibited by the peripheral cold receptors. Although sweating is primarily related to cranial temperature, the sensitivity of this relationship and in particular the threshold cranial temperature ('set point') at which sweating begins, is dictated by the skin temperature if this is less than 33°C.

Peripheral and central receptors show a very similar interaction in the control of thermoregulatory heat production (Fig. 5–2b). Thus shivering starts at a higher cranial temperature in a colder environment. As the

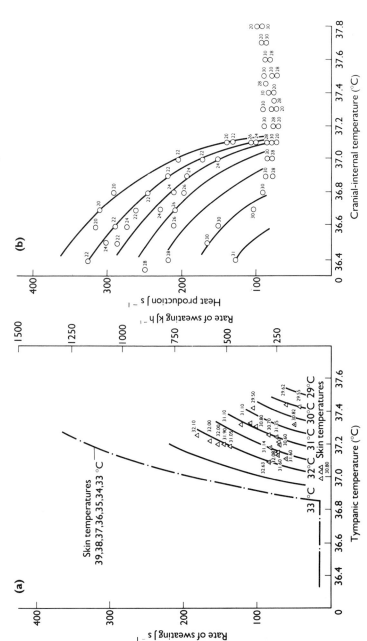

Fig. 5–2 (a) Inhibition of sweating in man by cold receptors in the skin (see text). Cranial temperature was monitored at the tympanic membrane. Skin temperature is indicated by △. (b) Inhibition of thermoregulatory heat production by central warm receptors (see text). Cranial temperature was monitored at the tympanic membrane. Skin temperature is indicated by ○. (From BENZINGER, 1969.)

environmental temperature increases, the cranial temperature must fall more before shivering starts.

The onset of sweating or shivering therefore depends primarily on cranial temperature, but cold receptors in the skin can inhibit the onset of sweating or promote the onset of shivering by influencing the 'set point' of cranial temperature at which these responses are switched on. Peripheral cold receptors thus influence the sensitivity of the central mechanisms in such a way as to modify the thermoregulatory responses in the light of environmental conditions.

5.4   The hypothalamus

The hypothalamus is the name given to the part of the brain below the thalamus which forms the floor and part of the lower lateral walls of the third ventricle (Fig. 5–3). Despite its small size it is very important in the internal regulation of body functions. Its influence on the anterior and posterior pituitary glands means that it governs much of the endocrine activity in the body, while it also has control of many aspects of

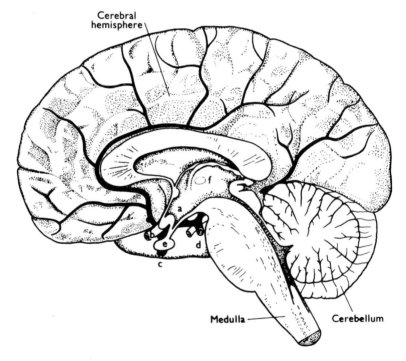

Fig. 5–3   Right-hand half of the human brain viewed from the mid-line. a, hypothalamus, forming wall of 3rd ventricle, f; b, optic nerve; c, internal carotid artery; d, basilar artery; e, pituitary gland.

autonomic nervous function. It plays an important role in regulating food and water intake, sexual behaviour and sleep and emotional responses, such as fear and anger. In addition to all these functions, it contains the structures ultimately responsible for regulating body temperature.

It has been said that attempting to investigate the functions of the various parts of the brain using the standard techniques of surgical interference, electrical stimulation and electrical recording, is like trying to deduce how a computer works, armed with a hatchet, a battery and a voltmeter! Nevertheless, notwithstanding the limitations of even the most sophisticated modern techniques, it has been possible to establish a great deal about the role of the hypothalamus in temperature regulation.

Surgical interference with the brain established in the early part of this century that the integrity of the hypothalamus was necessary for normal temperature regulation. Animals with the brain sectioned below the hypothalamus, or with the hypothalamus destroyed, became essentially poikilothermic.

A great deal is now known of the internal organization of the hypothalamic temperature regulatory centres as a result of experiments in which small areas have been destroyed, stimulated electrically or subjected to thermal stimulation.

5.4.1 Damage to the hypothalamic centres

During the past 30 years techniques have been developed which allow the accurate placement of fine electrodes into specific areas of the brain of anaesthetized animals. Once an electrode has been introduced into a particular part of the brain it can be used to stimulate it by means of a small electric current, or it can be used to monitor the electrical activity of the brain cells near its tip. In addition, such an electrode can be used as a means of destroying a small area of the brain. Such *focal lesions* occur around the uninsulated tip of the electrode when a large direct current is passed through it. The extent of the electrolytic damage to the brain can be graded by adjusting the intensity of the current and can be measured histologically at the end of the experiment.

Lesions placed in the anterior (front) part of the hypothalamus result in the inability of the animal to regulate its temperature in a warm environment, although it can still maintain a normal temperature if kept in a cold room (Fig. 5–4). The animal cannot activate its heat loss mechanisms: it cannot pant, sweat, or increase the blood flow to its skin. A similar inability to increase heat loss in a warm environment is occasionally seen in human patients suffering from disease or damage to the anterior hypothalamus.

Conversely, if lesions are placed in the posterior (rear) part of the hypothalamus, animals find difficulty in conserving heat and increasing heat production when exposed to the cold. However, their ability to lose heat in a warm environment remains unimpaired.

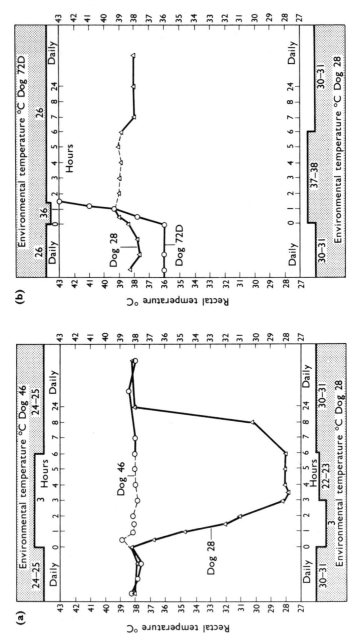

Fig. 5-4 Effect of hypothalamic lesions on temperature regulation in dogs. (a) Effect of a cold environment, (b) effect of a warm environment. Dog 46—unoperated; Dog 28—lesion in posterior hypothalamus; Dog 72D—lesion in anterior hypothalamus. Dotted lines indicate in (a) shivering, and in (b) panting. (From KELLER, A. D. (1950). *Phys. Ther. Rev.*, 30, 511.)

The conclusion reached from experiments involving localized destruction of parts of the hypothalamus was that there are two centres concerned with temperature regulation: an anterior 'heat loss' centre and a posterior 'heat production and conservation' centre.

5.4.2   Electrical stimulation of the hypothalamus

The above conclusion has received considerable support from the results of localized electrical stimulation of the hypothalamus. Stimulation of the anterior 'heat loss' centre results in vigorous panting, vasodilation in the skin and the inhibition of shivering. Such responses can be obtained even when the animal is in a cold environment. Stimulation of the posterior 'heat production and retention' centre causes shivering, vasoconstriction in the skin, piloerection and the inhibition of panting.

5.4.3   Thermal stimulation of the hypothalamus

Evidence from lesions and from electrical stimulation indicates that there are two anatomically separate, but reciprocally connected, regions of the hypothalamus concerned with the responses to hyperthermia and hypothermia. If this is indeed the case, then it remains to determine how hyperthermia activates the 'heat loss' centre and hypothermia activates the 'heat production and conservation' centre.

We have discussed previously the evidence for the existence of central thermoreceptors situated somewhere in those parts of the brain supplied by the internal carotid artery. Evidence will now be outlined which indicates that the thermoreceptors are in fact in the hypothalamus and, furthermore, that they are co-existent with the control centres described above.

A localized increase in brain temperature can be produced around the tip of an implanted electrode by diathermic heating (a.c. current of very high frequency). If such heating is applied to the 'heat loss' centre the animal manifests all the responses appropriate to an increase in heat loss. Furthermore, it has been shown that the response to heating is extremely sensitive; an increase of only $0.2–0.3°C$ produces a pronounced increase in heat loss. This heat loss is maintained, provided the hypothalamus is continuously heated, even though the core temperature is falling as a result of the increased heat loss.

Attempts to demonstrate that there are cold receptors in the posterior hypothalamus have produced equivocal results. Some investigators claim to be able to produce shivering, vasoconstriction in the skin and increased production of thyroid and adrenal medullary hormones when the hypothalamus is cooled. Other workers have not been able to obtain such responses and conclude that activation of the 'heat production and conservation' centre is primarily through impulses from cold receptors in the skin.

5.4.4   *Electrical recording from the hypothalamus*

It is possible to record the electrical activity of single cells in the brain by using very fine electrodes and a powerful amplifier. This technique has been used to determine whether there are cells in the hypothalamic temperature-regulating centres which alter their discharge frequency in response to changes in either hypothalamic or skin temperature.

Single-cell recording from the brain is technically very demanding and is a field in which great controversy has been generated due, for example, to the disparate effects of the various anaesthetics available. Nevertheless,

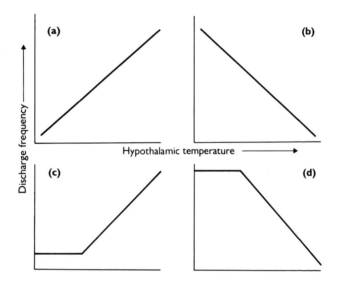

Fig. 5–5   Diagram to illustrate the most common patterns of temperature-related changes in the discharge rate of neurones in the anterior hypothalamus.

(a) Cells which *increase* their rate of discharge virtually linearly over the physiological range of local hypothalamic temperature (N.B. Q_{10} due to non-specific effects would be approximately 2, so specialized thermal sensitivity is not apparent unless the Q_{10} of the relation between temperature and activity is considerably greater than 2). These neurones are assumed to represent primary warm sensors.

(b) Cells which *decrease* their rate of firing virtually linearly over the range of normal hypothalamic temperature with a Q_{10} of less than 1. These cells occur infrequently and are assumed to be primary cold sensors.

(c) and (d) Cells which exhibit a non-linear relation between temperature and activity. In each there is a range of temperature insensitivity until a 'threshold' temperature is reached: thereafter activity increases (c) or decreases (d) linearly. These non-linear cells with a thermal 'set point' are believed to represent interneurones on the pathway to the effector systems, influenced by the primary sensors and by other factors.

studies during the last decade or so now point to the following general conclusions.

1. Temperature-sensitive neurones are found concentrated in the anterior hypothalamus: they also occur to a much lesser extent in the posterior hypothalamus and occasionally elsewhere. This supports other evidence indicating the unique delicate thermal sensitivity of the anterior hypothalamus (cf. section 5.4.3).

2. Temperature-sensitive neurones in the anterior hypothalamus can be separated into a number of categories depending upon the nature of their responses (see Fig. 5–5).

5.5   Models of thermoregulation

Now we have examined in some detail the characteristics of the hypothalamic areas involved in thermoregulation and, in particular, now we have information about the behaviour of individual neurones, it is possible to speculate about the possible organization of the neural 'thermostat'.

At its simplest we can envisage a system as shown in Fig. 5–6 where the bell-shaped temperature/activity curves of the primary hypothalamic warm and cold sensors intersect at the set point. In this model, first proposed by A. J. Vendrik in 1959, if the hypothalamic temperature comes to exceed the set point A, the warm sensors will be stimulated, while cold sensor activity declines – consequently heat loss is promoted while heat production is inhibited. Conversely, if hypothalamic temperature is below the set point, B, warm sensors and heat loss are depressed while cold sensors and heat production are augmented.

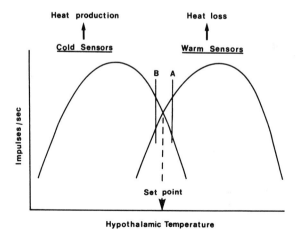

Fig. 5–6 A simple model to demonstrate a possible interaction of the temperature/ activity curves of 'cold' and 'warm' sensors in defining the set point.

Such a model is somewhat naïve and oversimplified because it ignores the fact that the hypothalamic set point is not constant; it can be varied by a number of factors such as skin temperature (section 5.3) or pyrogens (section 5.7.2). This complication can be embraced by the variable set-point hypothesis first proposed by H. T. Hammel and his colleagues in 1963. The principles of this hypothesis are illustrated in Fig. 5–7. The

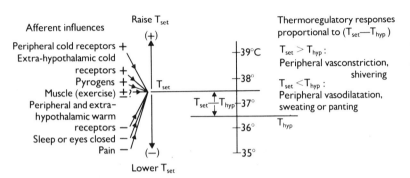

Fig. 5–7 The variable set-point hypothesis. Thermoregulatory effector responses are determined by the difference between actual hypothalamic temperature and 'required' hypothalamic temperatures as represented by the variable 'set point' which may be influenced by many afferent factors. (From BLIGH, 1966.)

basic idea is that the thermoregulatory effector responses are proportional to the error-signal generated by the difference between the set-point temperature (T_{set}) and the actual hypothalamic temperature (T_{hyp}). T_{set} can be influenced by a number of factors impinging upon the hypothalamus either neurally or in the blood, while T_{hyp} is, of course, dependent upon the overall balance between heat production and heat loss by the body. This basic hypothesis goes a long way towards explaining the neuronal basis of thermoregulation, but of course the above account represents only the tip of the iceberg of modern theory: an excellent detailed discussion can be found in BLIGH (1973).

5.6 Synaptic transmitters in the 'thermostat'

Most, if not all, neurones in the mammalian central nervous system appear to influence other neurones by the release of specific chemical transmitters at their synapses. During recent years a great deal has been learned about synaptic interconnections within the hypothalamic thermoregulatory areas from a study of their pharmacology. At first sight it might appear wildly optimistic to expect to produce defined thermoregulatory effects upon introducing a suspected transmitter artificially into even small regions of the hypothalamus. One might anticipate that such a procedure would elicit a multiplicity of non-specific side-effects. Surprisingly, however, such an approach has provided a valuable insight into the functioning of the thermoregulatory areas. Thus, injection of a transmitter into either the cerebral ventricles or the hypothalamus itself has demonstrated definite effects produced by noradrenaline (NA) and 5-hydroxytryptamine (5HT).

In cats, dogs and monkeys NA caused a fall in core temperature by promoting heat loss, while 5HT caused an elevation in core temperature by stimulating shivering and vasoconstriction in the skin. These effects were only seen when the transmitters were injected into the hypothalamus and not when injected elsewhere in the brain. To the great surprise of the investigators, however, some of the transmitters caused diametrically opposed effects when tried in certain other species. Thus in sheep, goats and rabbits, 5HT caused heat loss, while NA inhibited both heat production and heat loss. The reason for these species differences remains to be explained and it is certainly perplexing why such a fundamental homeostatic mechanism as thermoregulation should have evolved so dramatically differently.

RECAPITULATION

The mammalian hypothalamus contains two anatomically discrete areas concerned with temperature regulation (Fig. 5–8). The anterior 'heat loss' centre exerts control over superficial heat loss by conduction, convection and radiation by causing vasodilation in the skin and over evaporative heat loss by increasing sweating and thermal panting. If this centre is damaged the animal cannot protect itself from overheating. The centre is activated primarily by an increase in hypothalamic temperature but can be inhibited by impulses from the peripheral cold receptors, probably acting via the posterior 'heat production and conservation' centre. This posterior centre seems to be activated primarily by impulses from peripheral cold receptors, although there may also be hypothalamic receptors which respond actively to a fall in temperature. It functions reciprocally with the 'heat loss' centre and is inhibited when the latter centre becomes active in response to an elevation in hypothalamic temperature. The posterior centre governs heat production through shivering and the metabolic actions of thyroid hormones and adrenaline. It also conserves heat by reducing blood flow through the skin by piloerection and by the inhibition of sweating and thermal panting.

5.7   Fever (pyrexia)

We are all familiar with the idea that certain types of disease are associated with an increase in core temperature, but what is the mechanism of fever and what benefit, if any, does it convey upon the afflicted individual?

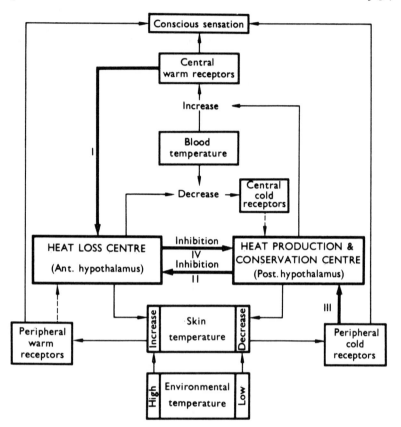

Fig. 5–8 Diagram to show the possible interrelationships between blood and environmental temperatures and the hypothalamic centres. Probable sources of conscious thermal sensation are also indicated. Important controlling pathways are indicated by dark lines. I: stimulation of anterior centre by hypothalamic warm receptors, II: inhibition of anterior centre by impulses from peripheral cold receptors acting via the posterior centre, III: stimulation of posterior centre by peripheral cold receptors and IV: inhibition of the posterior centre by central warm receptors acting via the anterior centre. Dashed lines indicate pathways of doubtful significance.

5.7.1 Pyrogens

Substances which cause fever are called pyrogens. In general there are two types: *exogenous pyrogen* (bacterial endotoxin) which is derived from microorganisms and *endogenous pyrogen* which is derived from the tissues of the animal itself. These two classes of pyrogen differ both chemically and in their action on the body.

Exogenous pyrogens are heat-stable high molecular weight polysaccharides particularly associated with Gram-negative bacteria. These pyrogens are extremely difficult to inactivate and can often survive normal sterilization procedures. Characteristically, the increase in body temperature they produce is biphasic; i.e. there are two distinct peaks of temperature. The first is believed to be due to the direct effect of exogenous pyrogen on the hypothalamus, while the delayed response is due to the stimulation of endogenous pyrogen (see below).

Endogenous pyrogen is a heat-labile protein which can be extracted from certain white blood cells (leucocytes). Administration of endogenous pyrogen causes a rapid monophasic pyrexia. It is believed that leucocytes contain a pyrogen precursor and that the pyrogenic stimulus (usually exogenous pyrogen) activates the cell allowing the production of endogenous pyrogen from the precursor: the endogenous pyrogen then effects the hypothalamus as described below.

5.7.2   *Pyrogens and the hypothalamus*

Pyrogens do not, as was once thought, disorganize the functioning of the hypothalamic thermoregulatory centres. In fact, it can easily be established that the febrile subject is using his thermoregulatory effector mechanisms in the appropriate way, but that the regulation is directed toward maintaining an abnormally high core temperature. In other words, the effect of pyrogens is to elevate the hypothalamic set point. This scheme provides an explanation for the sequence of events experienced during a febrile episode (Fig. 5–9).

The precise manner in which pyrogens increase the set point remains unclear. Despite intensive studies there is no convincing evidence to support an action via changes in NA or 5HT and indeed such an effect would seem *a priori* to be improbable in view of the comparable action of pyrogens in groups of species in which 5HT and NA have opposite effects on thermoregulation (see p. 45). It seems most likely that pyrogens affect the temperature/activity curves of the primary warm and cold sensors such as to shift the curves and thus the set-point intersection to a higher temperature (see Fig. 5–6 and imagine both the curves transposed to the right).

5.7.3   *Fever and survival*

The final questions we have to consider about fever concern its value to the febrile subject – does it in some way help the body's defence mechanisms to overcome the bacterial invasion, and if so, how? These basic questions raise a number of interesting subsidiary points.

Fever is a complex response whereby the body temperature is reset at a higher value and then actively *regulated* at the new level. Surely such a system would not have evolved unless it had some survival value? Against that one could point out that fever may work by elevating the body temperature to a point where the defence mechanisms are activated to a greater degree than is multiplication of the microorganism, but often this

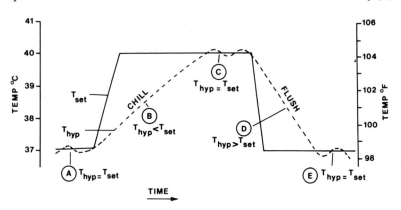

Fig. 5–9 Hypothetical scheme to illustrate the changes in set-point temperature (T_{set}) and hypothalamic temperature (T_{hyp}) during fever.

(**A**) Before fever begins, normal thermoregulation around normal brain temperature ($T_{hyp} = T_{set}$) – *normothermic.*

(**B**) Onset of fever. The set point has been elevated by pyrogen and therefore $T_{hyp} < T_{set}$ and the subject is *hypothermic.* In consequence, heat production is implemented; the patient feels chilled, shivers violently and intense vasoconstriction of the skin occurs.

(**C**) As a result of the increased heat production, the brain temperature is elevated to meet the increased set point and normal thermoregulation resumes – *normothermic.*

(**D**) The effect of the pyrogen subsides and the set point is reduced to normal. At this time the subject is effectively hyperthermic ($T_{hyp} > T_{set}$). This produces an acute sense of heat discomfort, sweating, vasodilation and panting, as a result of which body temperature decreases. This intense and dramatic increase in heat production used to be termed the 'crisis' and was regarded by the physician as indicative that the patient had 'turned the corner' and was on the way to recovery.

involves increasing the temperature to levels where hyperthermia, rather than the microorganism, may be the cause of death. In any event, if the above hypothesis were accepted then it would follow that the microorganisms by producing exogenous pyrogens were hastening their own destruction rather than that of the host (but is it in the interest of the microorganism to kill the host?). Finally, if fever *is* beneficial to the victim of infection, why does modern medicine routinely use antipyretics (drugs to reduce fever) such as aspirin?

Some insight into the possible value of fever to survival has come from recent fascinating new work on fever in lizards. The idea of a febrile lizard is, at first sight, improbable, since lizards are not homeothermic. However, you will recall from Chapter 2 that reptiles have well-developed behavioural thermoregulation and when presented with a graded thermal environment will position themselves so as to maintain a species-characteristic preferred or 'eccritic' temperature. It was found in the

U.S.A. in 1974, that if lizards were injected with bacterial pyrogen and placed in a temperature gradient, they positioned themselves so that they developed about 2°C of fever (i.e. their preferred temperature was 2°C above that before they were given the pyrogen). The presence of a febrile response in lizards immediately leads to the possibility of investigating the survival value of the fever, because of course the fever after pyrogen can be prevented by keeping the lizards in a constant temperature chamber. Figure 5–10 shows the results of such an experiment. It can be seen that

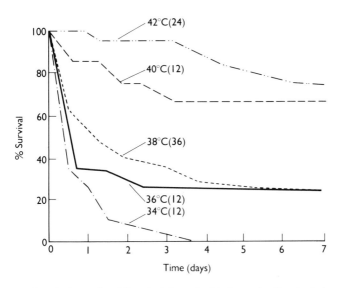

Fig. 5–10 Survival graph of lizards injected with bacteria. Survival time is a function of temperature. (Figures in brackets are the numbers of lizards in each group.) (Modified from KLUGER, M. J., RINGLER, D. H. and ANVER, M. R. (1975). *Science*, **188**, 166. Copyright 1975 by the American Association for the Advancement of Science.)

lizards allowed to develop a fever after injection with bacteria survived much better than those whose temperatures were maintained at lower values. Figure 5–11 indicates the probable explanation for the results. While the effectiveness of the lizards' antibacterial defence mechanisms can be assumed to be a linear function of temperature, the ability of the bacteria to reproduce (measured by the time they take to double their population) is essentially constant between 34–40°C, and above the latter temperature growth is actually depressed. It follows, therefore, that the higher the temperature the greater the probability that the lizards will successfully resist the bacteria.

On balance it seems that in mammals *moderate* fever probably assists in combating the invasion of microorganisms, since a small increase in core temperature would stimulate the defence mechanisms more than it would bacterial growth. However, since brain temperatures in excess of 43–44°C can themselves cause irreversible tissue damage and death, the use of antipyretics to control *extreme* fever seems justified.

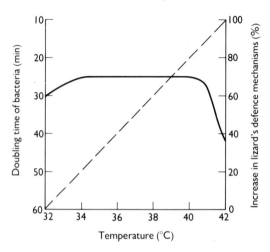

Fig. 5–11   *In vitro* doubling time of the bacteria (———) and theoretical increase in the lizards' defence mechanisms (– – – –) plotted against temperature. At 34°C, the growth of the bacteria far outweighs the lizards' ability to destroy these organisms, and all animals die. At 36 to 38°C, this difference decreases and the survival increases to 25%. Between 38 and 40°C, the lizards' defence mechanisms have improved so that a majority of the lizards survive the infection, even though the *in vitro* growth of the bacteria is unchanged from that in the range 34–38°C. At 42°C, the enhanced defence mechanisms together with a diminished bacterial growth rate result in essentially no deaths attributable to the bacterial infection. (Modified from KLUGER *et al.* (1975). Copyright 1975 by the American Association for the Advancement of Science.)

6 Adaptation to Hot Environments

6.1 Introduction

The core temperatures of most mammals and birds fall within the range 35–42°C and are, in consequence, higher than the air temperatures found in most terrestrial environments. It is not surprising, therefore, that much of the evolution of mammalian and avian temperature-regulation mechanisms has been in the direction of the efficient control of heat transfer from the body to the environment. Adaptation to a cold environment, therefore, merely entails a quantitative improvement in those mechanisms which restrict heat loss and which augment endogenous heat production, utilized by homeotherms in less demanding temperate environments. Therefore, given adequate food resources, effective adaptation to extreme cold has been evolved by mammals and birds from many orders.

On the other hand, efficient thermoregulation under conditions of extreme heat, particularly in environments where water is at a premium, is much more difficult. When the environmental temperature exceeds skin temperature, the normal thermal gradient is reversed and heat flows from the environment to the body. Such heat, together with the animal's own endogenous heat production, can be dissipated only by evaporation of water from the body. For this reason it is important to separate the physiological problems associated with a hot dry environment from those present in a hot humid environment.

In hot dry conditions the low atmospheric humidity facilitates evaporation, so that evaporative cooling can be extremely effective. However, since water is at a premium under these conditions, the extent of evaporative cooling is usually limited by the availability of water. In hot humid conditions on the other hand, there is no shortage of water, but the humidity of the atmosphere restricts evaporation and thus presents a limitation to evaporative cooling.

Relatively little study has been made of climatic physiology in hot humid environments, so our discussion will be restricted to the hot dry environment of the tropical desert, with special reference to the ways in which the camel and the kangaroo rat survive under these conditions.

6.1.1 Hot dry conditions

The most extreme hot dry conditions are found in tropical deserts which constitute an exceptionally unfavourable habitat for animal life. During the day, the heat from the sun is intense and vegetation provides little shade, while at sunset the temperature falls precipitously, even in summer, as heat is radiated into the cloudless sky. Over vast areas of desert there are no permanent surface water accumulations and years may elapse between rains.

It should be remembered when considering the gross meteorological characteristics of a particular environment, such as the desert, that these approximate only to the conditions experienced by large animals such as camels and ostriches. Meteorological data are not immediately or necessarily indicative of the true environmental conditions of smaller species. Most terrestrial vertebrates weigh less than 250 g and such creatures exist for much of their time in burrows, crevices or nests, in which the micro-climate is much more favourable than that of the world outside.

6.2   The camel

The camel is perhaps the most famous animal of the desert and its reputation rests mainly upon its ability to survive and indeed to travel for long periods of time without access to water. The problem of how this is accomplished has obsessed man's curiosity since earliest times, but only recently, mainly as a result of the work of K. Schmidt-Nielsen and his colleagues, has the mystery been unravelled. Schmidt-Nielsen directed his research towards answering four questions: Does the camel store water? Does the camel have an unusual core temperature and can it tolerate very high core temperatures? Does the fur act as an effective heat barrier? And finally, could the camel tolerate extreme dehydration?

6.2.1   Does the camel store water?

There is good evidence, to be discussed below, that camels do not overhydrate when allowed access to water. That is to say that the water they drink merely alleviates existing dehydration and restores the body fluids to their normal volumes. Thus, camels do not drink in *anticipation* of water requirements and cannot, therefore, be considered to store water. Nevertheless, it is instructive to examine some of the sites where water was previously thought to be stored.

Camels are ruminants and therefore have several chambers immediately preceding the true stomach. The largest of these, the rumen, differs anatomically from the rumen of typical ruminants and in the camel has bands of muscle which were at one time supposed to isolate certain sections of the rumen to allow them to function as 'water sacs'. The concept of 'water sacs' is a very old one. It was mentioned by Pliny and is still referred to today in some modern textbooks. However, it can now be discredited, because it has been established that the 'water sacs' cannot be effectively shut off from the rest of the rumen and in any case they are far too small to constitute a useful water reservoir. Furthermore, Schmidt-Nielsen has examined a number of freshly-slaughtered camels and did not find water in the sacs.

The hump has often been suggested as a site for water storage. Not liquid water of course, despite nursery stories to the contrary, but water produced from the oxidation of the fat stored in the hump. When 1 g of fat is oxidized it yields 1.07 g of water, so that a camel with a 40 kg hump comprising mostly adipose tissue might be thought to have the equivalent of over 40 l of water stored. The fallacy in this idea comes when the oxidation process is examined. Oxidation needs oxygen, which in turn requires the ventilation of the lungs. Air expired from the lungs is saturated with water vapour so, unfortunately, the process of obtaining the extra oxygen to oxidize the fat entails the loss of at least as much water vapour from the lungs as would be obtained by oxidizing the fat! Therefore, the fat in the hump does not constitute a water reservoir, although it does of course provide a valuable reservoir of metabolic energy.

To summarize: the camel does not store water: it does not drink in excess of its immediate requirements and in any case there are no anatomical sites where significant volumes of reserve water could be retained.

6.2.2 Does the camel have an unusual core temperature and can it tolerate very high core temperatures?

Part at least of the explanation of the camel's ability to live in the desert without requiring much water was revealed from a study of the animal's core temperature fluctuations.

When a camel has access to water its diurnal core temperature changes are much less than when it is deprived of drinking water (Fig. 6–1). In the dehydrated camel the maximum daytime core temperatures may exceed 40°C while the nocturnal core temperatures may approach 34°C. This represents a diurnal variation of 6°C.

The ability to tolerate large core temperature fluctuations means that the dehydrated camel can conserve water because, if the core temperature were to be kept constant, all the heat obtained from the environment during the day would have to be dissipated immediately by the evaporation of water. The fact that the camel stores heat during the day, by allowing its core temperature to increase, means that evaporative heat loss will be correspondingly reduced. For example, a 6°C temperature rise in a 500 kg camel of specific heat $3.35 \text{ J g}^{-1} \text{ K}^{-1}$ would result in the storage of about 10 000 kJ (2500 kcal). To dissipate this quantity of heat by evaporation would require more than 4 l of water. Instead, the camel loses the heat by conduction and radiation during the night. One incidental advantage of allowing the temperature to rise during the day is the reduction in the gradient for heat transfer from the environment to the camel.

It is not known whether the camel has an abnormally high lethal temperature, but since there have been no reports of core temperatures exceeding 41°C under normal field conditions it seems that the camel does not normally conserve water by allowing itself to overheat.

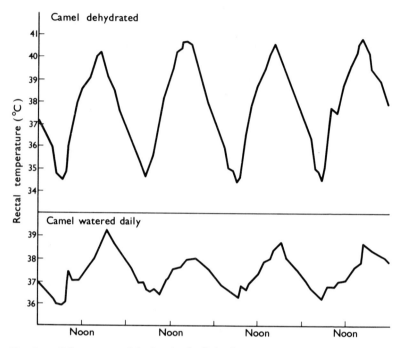

Fig. 6–1 When a camel is deprived of drinking-water its rectal temperature may show a diurnal fluctuation of as much as 5 or 6°C. When it is watered daily the fluctuations are much smaller. (From SCHMIDT-NIELSEN, 1964.)

6.2.3 Does the fur act as an effective heat barrier?

Surface insulation restricts the transfer of heat from the environment to the body so that the camel's fur might be expected to assist indirectly in the conservation of water if it retarded the heating up of the body. Observations by Schmidt-Nielsen indicate that the fur does indeed contribute to a large extent to the conservation of water. A shorn camel with less than 1 cm thickness of fur lost an average of 3 l of water per 100 kg body weight per day compared with a loss of 2 l per 100 kg per day in an unshorn animal with fur measuring between 3 and 14 cm in length.

6.2.4 Could the camel tolerate extreme dehydration?

Although the camel does not store water we have seen that it restricts its loss by evaporation by tolerating large diurnal variations in core temperature and acting as a heat store. Its fur also retards the rate of heat transfer to the body from the environment. Nevertheless, the animal still loses some water by evaporation from the skin and the respiratory tract, in the urine, although this can be highly concentrated and of small volume, and in the faeces, which are exceptionally dry. Since the camel does not over-hydrate it follows that the unavoidable water loss during a voyage across the desert must produce dehydration.

Perhaps the most significant factor in the ability of the camel to tolerate desert conditions is its ability to withstand dehydration. Other mammalian species, such as the dog and the rat, die under temperate conditions when dehydrated to a degree which produces a fall in body weight of between 12 and 14% of the original value. Death from dehydration occurs even sooner if such animals are dehydrated under conditions of heat stress. The camel, on the other hand, can readily withstand dehydration resulting in a 25–30% weight loss in summer desert conditions. One contributory factor to the camel's tolerance of dehydration is the fact that when dehydrated its plasma volume falls less than in other species. A fall in plasma volume causes a profound disturbance in the cardiovascular system and under hot conditions is particularly dangerous as it reduces the rate of transfer of heat from the core to the body surface.

The recovery from dehydration when water becomes available is particularly dramatic in the camel. A water deficit equivalent to 20% of the body weight can be redressed within 10 minutes. This entails drinking of the order of 70–100 l of water within that short space of time. It would be the equivalent of a 70 kg man drinking about 17 l (30 pints) of water. A camel can accurately rehydrate itself very rapidly after a period of dehydration while the process takes many hours in man.

6.3   The kangaroo rat (*Dipodomys*)

Just as the camel provides an extreme example of the functional adaptation of a large animal to desert life, the kangaroo rat serves as an extreme representative of small mammals. The camel is tolerant of high temperatures and has survived by developing physiological mechanisms which enable it to cut down its water requirements and to survive extreme dehydration. The kangaroo rat, on the other hand, is not particularly tolerant of high environmental temperatures and goes to great lengths to avoid such conditions: it is nocturnal and it avoids the heat of the day by retreating into deep burrows.

The main physiological interest in the kangaroo rat has centred around the way it obtains water and its apparent ability to survive without drinking. This remarkable adaptation enables the animal to flourish in extremely arid conditions where there is neither surface water nor juicy plants and bulbs. The kangaroo rat can survive on air-dried food because its water economy is so efficient that its requirements can be met by the small quantity of free water in its diet of dry seeds and other dry plant

material together with the metabolic water released when dietary
components are oxidized.

Dipodomys can therefore survive without drinking, and so can inhabit
areas where there is no surface water. It limits its water loss by living in

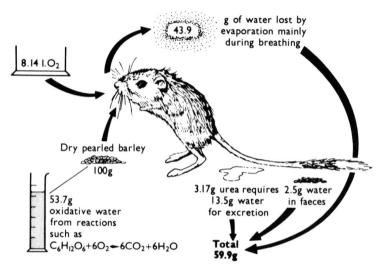

43.9

8.14 l.O$_2$

g of water lost by
evaporation mainly
during breathing

Dry pearled barley
100g

53.7g
oxidative water
from reactions
such as
$C_6H_{12}O_6 + 6O_2 \rightarrow 6CO_2 + 6H_2O$

3.17g urea requires 2.5g water
13.5g water   in faeces
for excretion

Total
59.9g

Fig. 6–2 The water relations of *Dipodomys*, the kangaroo rat, when fed on dry
pearled barley. The difference of 6.2 g between the total water loss of 59.9 g and
the water derived from oxidation of food is made good if the pearled barley is
in equilibrium with air of 20% relative humidity. (From CHAPMAN, G. (1967).
The Body Fluids and their Functions. Studies in Biology no. 8. Edward Arnold,
London. Based on SCHMIDT-NIELSEN, 1964.)

relatively cool burrows below the hot surface of the desert and by
emerging for food only at night. Sufficient water is obtained from the
oxidation of food material and from the free water absorbed in the food.

6.4 Desert birds

Birds living in the desert might seem at first sight to labour under at
least two disadvantages when compared with mammals of comparable
size. First, they are non-fossorial; they cannot retreat into burrows and
escape extremes of temperature, and secondly, their feeding habits and
their poorly developed sense of smell mean that they are necessarily active
during daylight hours. On the other hand, birds have the ability to fly and
this greatly extends their mobility and their range from sources of water
and may also enable them to lessen heat stress by escaping to high
altitudes.

Although birds do not construct burrows, they minimize heat stress by restricting their feeding activity to those cooler parts of the day just after sunrise and just before sunset. During the heat of the day, small birds escape to the shade and remain virtually motionless. On the other hand, large birds do not generally rest in the shade for long periods: these individuals spend the day circling high in the air. They may attain altitudes of up to 1000 m, presumably rising with little effort in thermals, and thereby remain, by virtue of the altitude, in air at a temperature about 10°C lower than the ground temperature.

6.4.1   Temperature regulation

Body core temperature in birds is generally within the range of 40–42°C and is thus several degrees higher than that of most mammals. In addition, birds can tolerate high core temperatures rather better than mammals: their lethal temperature is in the range 45–47°C. Desert birds do not seem to be exceptional in terms either of normal core temperature or of lethal temperature when compared with non-desert birds. Nevertheless, they are at an advantage when compared with desert mammals, because their higher core temperature and greater tolerance to hyperthermia both diminish the potential heat load and thus decrease the water requirement for evaporative heat loss. The high resting body temperature, and the further rise in body temperature produced in birds by exposure to high environmental temperatures, may both be considered an adaptive factor favouring heat dissipation by conduction and radiation and the resultant conservation of water.

6.4.2   Water balance in desert birds

As we have just discussed, the need for evaporative heat loss is minimized both by avoidance of extreme environmental heat stress and by the high core temperature. Nonetheless, at core temperatures of about 42°C evaporative heat loss through the respiratory tract becomes necessary and is mediated by the onset of extremely rapid shallow panting. In high environmental temperatures this thermoregulatory evaporative loss of water constitutes by far the major route of water loss, especially when it is remembered that there are no sweat glands and that very little water is lost in the urine since uric acid is the end product of protein metabolism in the bird. Evaporative heat loss may increase five-fold when a bird is transferred from a temperate environment to an extremely hot environment and in the latter case may represent a daily water loss equivalent to 23–30% of the body weight.

Water obtained from the oxidation of food has been shown in all cases studied to be inadequate to satisfy alone the requirements of desert birds. It follows from this that they must obtain liquid water, either by drinking, or from moisture in the diet. Thus most desert birds must remain in the proximity of a watering place, although the ability to fly means that the term proximity may be considered to embrace areas many miles from water. It is of interest to note that the male sand-grouse (*Pterocles*) carries liquid water to the chicks by saturating its ventral feathers with water when drinking. The young suck the water from the feathers on his return.

6.5   Man in the desert

Deserts are no longer shunned by civilized man; they are increasingly exploited for their mineral resources (particularly oil), for agriculture and in time of war. Man's ability to live and work under desert conditions can be attributed to his technology rather than to his physiology. For, as in other mammalian species, man's thermoregulation during the desert day depends upon the evaporative loss of heat and to sustain this process he needs an adequate water intake. To this end, technology helps to ensure a supply of water in many ways, ranging from the basic development of portable water storage units such as goat-skin bags and the domestication of pack animals to carry them, right up to modern water pipelines, deep wells and even jet aircraft dropping supplies. Technology also assists desert survival in a second way, namely by enabling man to lessen the rigours of the climate through the development of clothing and the construction of shelters.

We shall not be concerned further with the technology of survival but rather with some of the physiological responses of man during heat stress. Much of the knowledge in this field is derived from the classic studies of E. F. Adolph and his co-workers on U.S. army personnel during the Second World War.

6.5.1   Sweating

As described in Chapter 4, the volume of sweat secreted during heat stress can be surprisingly large and can be related both to the air temperature and to the physical work performed (Fig. 6–3). It can be seen that under certain conditions a water loss in excess of 1 l/h can take place and in extreme circumstances losses in excess of 4 l/h have been observed for short periods. Such high rates of sweating cannot, of course, be maintained in the absence of adequate water intake. In addition to water, sweat contains a variable amount of sodium chloride but always enough to represent a serious salt loss during severe sweating. Thus the consumption of salt-free water after severe sweating does raise another problem: the dilution of the body fluids. This manifests itself as severe muscular cramps – 'heat cramps'. Such cramps had long been associated with colliers working in deep mines and with stokers in steamships and were attributed to the effects of heat. It remained for J. S. Haldane to relate the cramps to the salt lost during severe sweating and to suggest the simple effective remedy of adding NaCl to the drinking water.

6.5.2   Dehydration

The most copious sweat rates dictated by hot environments can be sustained without harm, provided that the water and salt intake is

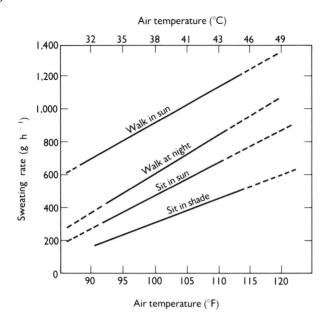

Fig. 6–3   Rates of evaporative heat loss in the desert., All subjects wore light tropical uniforms. Walking was at about 3–3.5 m.p.h. for 50 min in each hour. Solid lines are pooled experimental results from several subjects. Dotted lines are tentative extrapolations beyond experimental data. (From ADOLPH, E. F. (1947). *Physiology of Man in the Desert.* Interscience, New York.)

adequate, but what are the consequences of an inadequate salt and water intake under these circumstances?

Experimental dehydration of human subjects cannot of course be continued to dangerous levels, but data from experimental work in man, together with results from animal experiments and occasional observations on human subjects rescued from the desert suffering from very severe dehydration, have enabled a fairly comprehensive picture of progressive dehydration to be compiled.

The complex changes during dehydration cannot be discussed here, but it is important to appreciate one surprising fact. Even during the most severe dehydration, the rate of sweating seems to be dictated by the need to lose heat rather than by the need to conserve water. Thus the water loss due to sweating continues unabated even though the dehydration has severely decreased the plasma volume. This decrease in plasma volume results in an increase in the viscosity of the blood, placing a great strain on the heart which thus becomes less able to maintain an adequate blood flow through the skin. As skin blood-flow falls, the transfer of metabolic heat from the core of the body to the skin is impaired. An explosive increase in core temperature to the lethal limit of 41–42°C ensues and the victim dies, despite the fact that the skin is still relatively cool due to continued sweating.

6.6   The carotid rete

In some species, such as the sheep, goat and cat, the arrangement of the arterial supply to the brain is such that the blood in the carotid arteries is precooled in a complex rete (net) by the counter-current heat exchange with cool venous blood draining the nasal cavity (Fig. 6–4). Cooling the carotid arterial blood in the rete before it enters the Circle of Willis and supplies the brain, means that the main mass of the body can become hyperthermic by several degrees without this producing an equivalent and prejudicial increase in brain temperature: in other words brain temperature is to a limited extent divorced from body temperature. It would be expected, and there is some evidence for it, that species with a carotid rete would have a greater tolerance of heat stress than species like man, where the carotid passes through the cavernous sinus as a single vessel in which little heat exchange occurs.

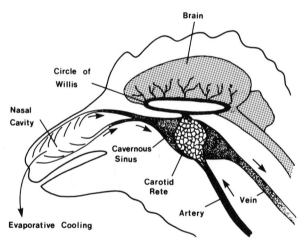

Fig. 6–4   The carotid rete (see text).

7 Adaptation to Cold Environments

7.1 Introduction

Climates which include periods of prolonged and severe cold promote one of three responses in an animal species. It may avoid the cold by moving to a region with a more favourable climate: this may be simply a seasonal migration or it may involve a long-term drift of population to warmer regions. Secondly, the species may adapt itself in such a way that it can tolerate the cold, undergo a normal life cycle and exploit the ecological potential of the environment. Finally, a species may compromise with the environment, living an active life and reproducing during the more favourable period of the year while spending the rest of the year in an inactive, hibernating state.

Discussion of migration or population fluxes does not fall within the scope of this book, so that this chapter will be concerned with adaptations to active life in the extreme cold and with mammalian hibernation.

7.2 Poikilotherms in the cold

In Chapter 2 we saw that moderate cooling of both invertebrate and vertebrate poikilotherms may induce physiological changes which enable the organisms to function more efficiently at lower temperatures (acclimatization). As cooling proceeds, however, the animal is eventually faced with the problem of avoiding freezing. There are two ways in which this can be achieved. First, the freezing point of the animal's body fluids can be lowered by the addition of solute and secondly the animal may 'supercool'.

7.2.1 'Antifreeze' in insects

Many insects habitually exposed to extreme cold lower the freezing point of their body fluids by the secretion of glycerol. The most extreme example is found in *Bracon cephi*, an insect parasite of the Canadian wheat stem sawfly. Larvae of the parasite can survive prolonged exposure to temperatures as low as −40°C. This is in part due to supercooling (see below), but can also be attributed to the fact that the glycerol concentration in the haemolymph may approach 5 M, which will lower the freezing point to −17.5°C. The use of glycerol as an 'antifreeze' by certain insects foreshadows man's use of similar compounds to protect the cooling system of cars during winter.

In addition to lowering the freezing point of the body fluids, glycerol also tends to protect the tissues from the damaging effect of freezing should this eventually occur. It probably achieves this protection both by ensuring that the increase in electrolyte concentration resulting from ice formation does not readily reach toxic levels and by its high viscosity, which tends to decrease the rate of formation of ice crystals and to change their pattern.

7.2.2   *Supercooling*

Under exceptionally favourable conditions ordinary water can be progressively cooled to −41°C before ice crystals appear. This phenomenon is called '*supercooling*'. As the temperature gradually falls below the normal freezing point (0°C) the molecular arrangement becomes more and more ice-like: the individual molecules slow down and move closer together and it becomes more likely that they will approach an ice-lattice structure. Eventually a crystal nucleus of ice is formed which grows until finally the entire water mass freezes. The focus for initial ice formation may be an aggregation of water molecules or freezing may be initiated round a foreign substance which is said to 'seed' ice formation.

Terrestrial arthropods during hibernation can supercool to levels of 25–30°C below the freezing point of the body fluids, which may itself have already been lowered by glycerol. Thus *Bracon cephi* can survive a temperature of −47°C without freezing, due to a 17°C lowering of the freezing point by glycerol and 30°C supercooling beyond this point.

The metastable state of supercooling is not harmful *per se* to the insect, but of course the metabolic slowing down, experienced during cooling down to the freezing point, continues further as the animal supercools. Supercooled insects are thus inactive and can be considered to be in a state of dormancy during the overwintering period.

SUPERCOOLING IN FISH   Fish do not hibernate and remain relatively active even when supercooled. That fish can supercool was demonstrated by Scholander and his colleagues as a result of studies on teleost fish in arctic fiords. Blood from fish taken from deep water had a freezing point between −0.9° and −1.0°C (the freezing point was taken as the temperature at which a small ice crystal was in equilibrium with unfrozen blood). Since the water temperature at the depth from which the fish were taken was −1.73°C it followed that the fish were supercooled by almost 1°C. The supercooled state is an unstable one and supercooled fish in supercooled water at −3°C could be frozen by seeding the water with ice (Fig. 7–1).

It follows from these observations that supercooled fish are safe provided they do not approach the surface ice. Thus the survival of *deep-water* fish in water at −3°C can be attributed to supercooling.

Supercooling does not provide an explanation for the ability of other species of teleost fish to swim just under the ice, for stable supercooling would not be possible in close proximity to ice. Protection from freezing in shallow-water fish swimming in sub-zero arctic seas seems to result from an increase in the osmolarity of their blood plasma. The consequent

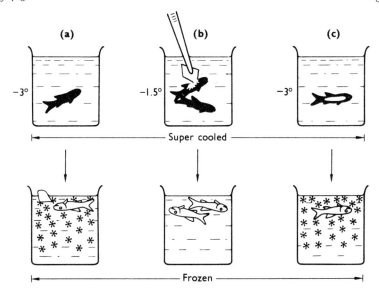

Fig. 7–1 Seeding experiments on *Fundulus* (temperature in °C). (a) Fish freezes when supercooled water is seeded. (b) Supercooled fish freezes when touched with ice in water above freezing point. One fish may, by contact, seed another. (c) Thawed fish with ice core freezes, and seeds supercooled water. (From SCHOLANDER, P. F. *et al.* (1957). *J. cell. comp. Physiol.*, **49**, 5.)

lowering of the plasma freezing point of these fish is such that it approaches the value for sea water. The increase in plasma osmolarity is partially attributable to chlorides but the other components responsible have not yet been identified: no glycerol is present.

7.3 Non-hibernating homeotherms

Non-hibernating homeotherms maintain their core temperature under conditions of extreme cold predominantly by the increased efficiency of their insulative mechanisms. It should be noted, however, that although core temperature is maintained, the temperature of the extremities may fall to extremely low levels (Fig. 7–2). The fact that their temperature approaches that of the environment means that the poorly insulated extremities do not constitute a major source of heat loss.

Nevertheless, the tissues in these regions must still function at these low temperatures, and in fact they exhibit some strikingly different properties when compared with similar tissues in the warmer regions of the body. For example, nerves show greater resistance to cold in the extremities than in the core. In herring gulls acclimatized to cold conditions, the tibial nerve, which runs from the spinal cord to the muscles in the foot,

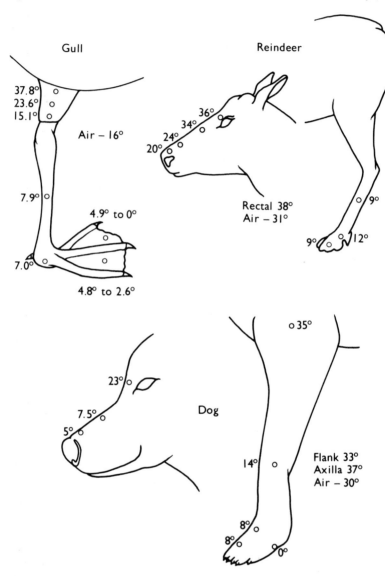

Fig. 7–2 Temperatures at extremities of arctic animals are far lower than the internal body temperature of about 38°C, as shown by measurements made on Eskimo dogs, reindeer and sea gulls. Some extremities approach the outside temperature. (From IRVING, L. (1964). Terrestrial animals in cold; birds and mammals, pp. 343–48 in *Adaptation to the Environment*, DILL, D. B., ADOLPH, E. F. and WILBER, C. G. (Eds.), in: The American Physiological Society's series 'Handbook of Physiology'.)

shows regional differences in its ability to conduct impulses in the cold. The section which is normally kept relatively warm by the feathers on the upper leg, is blocked by cold at 11.7°C whereas the nerve from the naked lower leg conducts impulses down to 2.8°C. Thus different parts of the same nerve cell have different temperature sensitivities.

There may also be other changes in the properties of tissues exposed to cold. Fat from the distal parts of the legs of arctic animals such as the caribou, has a melting point more than 30°C below that of fat in the core, while the fat in the hooves melts at 0°C or less and thus presumably allows the hooves to remain flexible and supple at these temperatures.

7.3.1   Insulation

Mammals can decrease heat loss from their surface by depositing a layer of subcutaneous fat and by developing an efficient air-trapping surface coat of fur. Aquatic mammals such as whales are limited to the first method of insulation and reduce heat loss very effectively by their thick layer of blubber.

The fur of terrestrial arctic animals constitutes an extremely effective thermal barrier which can perhaps best be appreciated by comparison with the thermal barrier provided by human clothing. A unit has been developed for this purpose. It is the 'clo' unit, which equals the insulation provided by the clothing which would be worn at a room temperature of

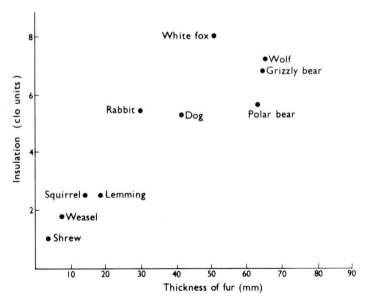

Fig. 7–3 Insulating capacity of fur from various animals; 'clo' units, see text. (Modified from SCHOLANDER, P. F. *et al.* (1950). *Biol. Bull. Mar. biol. Lab., Woods Hole*, **99**, 225–36.)

20°C (i.e. underclothes, shirt, jacket and trousers). It may seem a rather arbitrary and unscientific unit but it has gained wide acceptance and allows a ready appreciation of the effectiveness of thermal insulation. Figure 7–3 shows the insulative capacity of fur from various species. In considering this data it is worth remembering that the maximum practical insulation to be gained by arctic survival clothing is about 6 clo units. (If more clothes are worn, movement is severely impaired.)

7.3.2   Metabolic response to cold

When the regulation of heat production was discussed in Chapter 4 it was noted that there was an increase in heat production in a naked human as soon as the environmental temperature fell below 28°C. In arctic animals the highly efficient insulation means that heat production need not increase until the environmental temperature falls to much lower levels. This is illustrated in Fig. 7–4 which shows the environmental temperatures at which heat production begins to exceed basal levels in

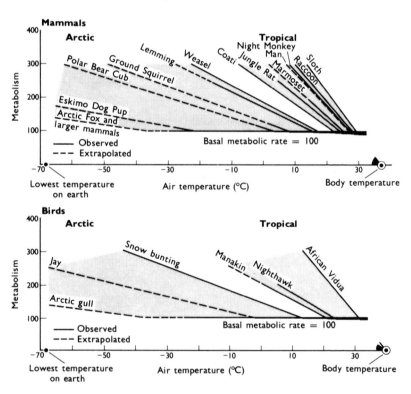

Fig. 7–4   Heat regulation and temperature sensitivity in (above) arctic and tropical mammals and (below) arctic and tropical birds. (From SCHOLANDER et al., 1950.)

representative tropical and arctic mammals. It can be seen that relatively large arctic species, such as the white fox, do not even begin to increase their heat production until the environmental temperature reaches about −40°C and need but a small increase in metabolic rate to survive in the coldest conditions on earth. Thus arctic foxes can sleep on open snow at −80°C for an hour without any fall in core temperature. The differences in metabolic responses to cold in arctic and tropical mammals are also seen in birds (Fig. 7–4).

7.3.3   The problems of small animals

Small mammals have a large surface area : volume ratio. As J. S. Haldane pointed out, 5000 mice weigh as much as one man but have seventeen times his surface area. Surface area determines heat loss, so that small animals lose more heat per unit body weight than large animals and must therefore eat proportionately more in order to support the increased heat production necessary to compensate for the heat loss. It follows from this that it is thermodynamically advantageous for a homeotherm to be large, particularly in colder environments and when food is scarce. There is, in consequence, a tendency for the members of a given species to be larger in the colder parts of its ecological range. This tendency is known as Bermann's Rule.

A further disadvantage suffered by small animals is that the length of their fur is limited by their small size. Many species compensate to some degree for the deficiencies of their own insulation by burrowing into the snow or substratum and thus creating a warm micro-climate of their own. A few birds such as the ptarmigan (Lagopus) also burrow, but the more usual behavioural response of birds to adverse climatic conditions is migratory flight.

7.4   Adaptive hypothermia

7.4.1   Hibernating mammals

Hibernation (winterschlaf, winter dormancy) may be defined as a periodic phenomenon in which body temperature falls to a low level, approximating to the ambient, and in which heart rate, metabolic rate and other physiological functions fall to correspondingly minimal levels. Spontaneous arousal is possible at any time during hibernation.

Hibernation is an adaptation to overcome periods of climatic stress associated with food shortage and is practised by certain species within three mammalian orders: Rodentia, Insectivora, Chiroptera. Hibernators are homeothermic during the summer, although the precision of temperature regulation is less than in non-hibernators.

During the winter, when food is scarce, the low environmental temperature would require a higher metabolic rate to maintain core temperature; under such conditions, the hibernator becomes virtually poikilothermic. It spends its time in an inactive, torpid state and supports its much reduced metabolic requirements by the energy stores it laid down in preparation for hibernation.

THE HIBERNATING STATE   During hibernation, the metabolic rate falls to as little as 1/70 of the basal non-hibernating rate and the core temperature falls to within 1–2°C of the environmental temperature (it should be noted that hibernators tend to retreat into burrows or nests in which the micro-climate is somewhat more favourable than in the outside environment).

If the local environmental temperature falls to dangerously low levels, where there would be danger of the tissues freezing, hibernators have two safety mechanisms which tend to prevent this occurrence: the metabolic rate may increase without the animal waking up, a relic of homeothermy, or the animal may undergo complete arousal.

The heart rate during hibernation may fall to as low as 5–6 beats per minute, but, despite this, intense vasoconstriction together with an increase in the viscosity of the blood ensures that the blood pressure remains adequate. The respiratory rate may decline to one respiration per minute and kidney function is much reduced. There is still formation of some urine and the animal may undergo periodic arousal for purposes of micturition. There are profound changes in the endocrine system. Many glands, such as the thyroid, adrenals, pituitary and gonads show signs of regression and inactivity, and cold no longer stimulates the thyroid and adrenal glands. However, despite extensive research on the endocrine system, it is still not clear what is the precise significance of specific endocrine changes to the induction and maintenance of the hibernating state.

ENTRY INTO HIBERNATION   There are two main patterns for entering into hibernation. Some species, such as the ground squirrel (*Citellus beecheyi*), enter hibernation gradually over several days. The nocturnal core temperature falls progressively lower over several successive nights in a series of 'test drops' until a critical point is reached where the temperature fall becomes non-reversible and the animal enters prolonged hibernation. In other species such as the woodchuck (*Arctomys* spp.), there is a single relatively rapid entry into hibernation which occupies only a few hours.

The trigger for hibernation is not well understood but it is clearly related to environmental temperature, lighting regime and the extent of the food stores deposited in the body. Some interesting work has indicated that a blood-borne substance may trigger hibernation in certain species. Thus the injection of a small volume of blood taken from a hibernating ground squirrel into a non-hibernating individual induced hibernation within 48 hours. Such hibernation could be induced in March, when ground squirrels are not normally prepared for hibernation, and was maintained throughout the summer. The factor, or factors, in the blood responsible for inducing hibernation have not yet been characterized.

AROUSAL FROM HIBERNATION   Although most hibernating species remain in their burrows for many months, most awaken from dormancy several times during this period. These periods of arousal may last a few hours or several days. During this time metabolic waste products are eliminated and some species, such as the dormouse, may eat some of the food previously stored in the burrow.

Arousal is a complex and highly coordinated process. There is a sudden increase in heat production and oxygen consumption which can be almost entirely attributed to shivering, to the activity of the heart and to the metabolism of brown fat. The heat produced is distributed via the circulation to particular organs in turn. The heart, lungs and brain, together with other less vital structures in the head and thorax, are warmed before the abdomen and extremities. This circulatory co-ordination is organized by the action of vasoconstrictor nerve fibres of the sympathetic system and means that in early arousal there may be as much as 20°C difference in temperature between the thorax and abdomen. Changes in heart and brown fat temperature and rectal temperature are illustrated in Fig. 7–5.

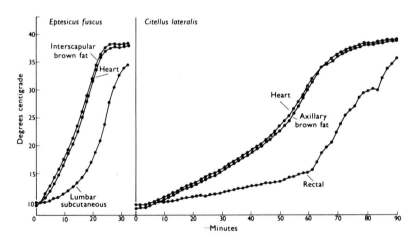

Fig. 7–5 Patterns of body temperature during arousal in the big brown bat (*Epesicus fuscus*) and the golden-mantled ground squirrel (*Citellus lateralis*). Environmental temperature 6–8°C. (From HAYWARD, J. S. *et al.* (1965). *Ann. N.Y. Acad. Sci.*, **131**, 441.)

7.4.2   Daily torpor

In addition to seasonal hibernation discussed above, there exists in certain species a daily cycle of intense activity alternating with periods of profound torpor. This behaviour is characteristic of small birds and mammals which have very high metabolic rates. The situation is particularly acute in small insectivorous bats (*Microchiroptera*) and ·in humming-birds (*Apodiformes*) which because of their aerial habits cannot carry large energy stores and are further restricted by their feeding (day-time in humming-birds and night-time in insectivorous bats). In both these groups the body temperature and the metabolic rate is high when active and feeding, but both fall precipitously when feeding ceases.

7.4.3   Dormancy in poikilotherms

Many poikilotherms overwinter in a state of dormancy with a much decreased metabolic rate and a body temperature very close to that of the environment. This is not comparable with true hibernation as defined on p. 67, first because arousal is not possible until the animal has been *passively* warmed, and secondly, there is no 'fail-safe' device as in mammalian hibernators, who will increase their metabolism or wake up if the body temperature approaches freezing point. Under such conditions poikilotherms freeze.

8 Cryobiology

8.1 Introduction

Cryobiology is the name given to the study of biological phenomena at low temperatures. During the last two decades it has provided some of the most exciting advances in the whole of biology, many of which have already found far-reaching practical applications.

Much cryobiological research is directed toward the ultimate extrapolation of the principle that cold depresses metabolic activity; i.e. that all metabolism in a cell, tissue or organism should effectively cease if the temperature is lowered sufficiently. Under these circumstances the processes of ageing and deterioration cease and the tissue could be preserved for ever in a state of 'suspended animation'. The implications of such a concept are immense, for it opens the way, in theory at least, to the long-term preservation of living material. The ability to store sperm for long periods without apparent deterioration has already had profound effects on animal husbandry and is also finding applications in some forms of infertility in man. It is now also possible to store certain organized tissues such as corneas, bone fragments, skin and heart valves at low temperatures without deterioration so that 'banks' of such material can be set up to aid surgical grafting and replacement procedures. However, the problem of freezing and preserving large tissue masses such as kidneys, hearts, livers and ultimately, of course, the entire mammalian organism, is proving much more intractable.

8.2 General problems in freezing tissues

It is necessary to lower the temperature to well below 0°C in order to reduce the metabolism of living tissues sufficiently to make long-term storage possible. Unfortunately, living tissues contain a high proportion of water which freezes at sub-zero temperatures and ice formation in tissues tends to cause irreversible damage. The main problem in cryobiology therefore is to devise means of cooling tissues to very low temperatures without causing the damage normally associated with the crystallization of water. Two general approaches to this problem have been made: first, attempts to reduce or abolish the harmful effects of ice formation, and secondly, attempts to preclude ice formation by removing the water before or during freezing: 'freeze-drying'.

8.2.1 Prevention of freezing injury

In order to understand the techniques devised to reduce tissue damage during freezing, it is necessary to outline briefly the several mechanisms by which such damage may possibly be caused: (1) mechanical damage to cell components during ice crystal growth; (2) denaturation of cell proteins due to electrolyte concentration, as solvent water is incorporated into ice crystals; (3) dehydration sufficient to precipitate proteins from solution; and (4) the removal of structurally important water molecules. A great deal of work has been devoted to an analysis of the relative contribution of these and other mechanisms to the damage caused by freezing, but the process is still little understood. Nevertheless, a number of techniques have been devised to reduce or prevent freezing injury.

PROTECTIVE ADDITIVES   We have discussed previously (Chapter 7) the fact that certain insects can survive in sub-zero temperatures by a combination of supercooling and depression of the freezing point of the body fluids by the secretion of glycerol. It is therefore interesting to record that the addition of glycerol to the experimental medium can convey some protection to isolated cells during freezing.

Glycerol may protect living cells against injury during freezing and thawing in several ways. These include lowering the freezing point of the medium, changing the crystalline pattern of extracellular ice formation during freezing and buffering the salts in the residual external medium once the water there has begun to crystallize. A similar buffering action within the cell may also be important since human red blood cells, which are permeable to glycerol, are protected from the effects of freezing down to at least $-79°C$ (the temperature of solid CO_2), whereas bovine red blood cells, which glycerol cannot enter, are destroyed. The internal glycerol probably protects the human red blood cell against the rise in internal K^+ concentration as ice forms. The protective action of glycerol on spermatozoa does not, however, depend upon penetration of the cell membrane and probably results from the ability of glycerol to prevent the external salt concentration from reaching damaging levels.

A number of other additives have been developed since the original work on glycerol in the 1950s and of these dimethyl sulphoxide is perhaps the most promising. It is a normal product of mammalian lipid metabolism and has the advantage that it tends to pass through most cell membranes more readily than glycerol.

RATE OF FREEZING   Much research has been devoted to the effects of the rate of freezing on the subsequent survival of frozen cells and tissues and it has been found that there are marked variations in the responses of different cell types. However, two general principles have been established on the basis of existing data: (1) where biochemical changes are the main source of cell injury, the extent of such injury depends on the degree of dehydration and on the temperature, so that faster cooling rates provide less time for such injury to occur; and (2) where intracellular crystallization is the prime source of damage, this can be minimized by a slow rate of freezing. Thus, the optimal rate of freezing for any particular tissue will be one at which injury is minimal and will depend upon the relative contribution of biochemical changes and intracellular ice crystal formation to such injury.

RATE OF THAWING   The rate of thawing seems generally to have much less effect than the rate of freezing on the survival of frozen tissues. Thus, if a tissue was frozen slowly it seems to matter little whether it is thawed rapidly or slowly. However, if a tissue was frozen quickly in order to minimize biochemical damage it is essential that it should be thawed rapidly. Since the avoidance of biochemical injury during freezing required a rapid passage through the intermediate sub-freezing temperatures where reaction rate was appreciable, it is not surprising that this temperature zone must also be passed through rapidly during thawing.

STORAGE TEMPERATURE   While many frozen materials appear stable at the temperature of solid CO_2 ($-79°C$), others do not. Red blood cells stored at this temperature for 40 days show a considerable decrease in survival, whereas storage at $-92°C$ results in no loss. Other tissues show some deterioration when stored for long periods even at $-92°C$. However, since reaction rates show an exponential relation with temperature, it would seem theoretically probable that storage at the temperature of liquid nitrogen ($-197°C$) would effectively eliminate deterioration, and indeed accumulating practical experience shows that this is the case. This implies that, in the absence of damage caused by the processes of freezing and thawing, living tissues could be stored indefinitely at $-197°C$ without deterioration.

8.3   Low temperature storage

The principle of prolonged storage at very low temperatures seems then to be a sound one, provided always that the transitional stages of freezing and thawing can be accomplished without damage. It remains to describe to what degree this has so far been achieved in practice.

8.3.1   Freezing single cells

Many microorganisms survive freezing conditions in nature as spores which have an extremely low water content. Such spores are also very resistant to freezing under a variety of laboratory conditions, presumably again as a consequence of their dehydrated state. Hydrated microorganisms often show remarkable ability to withstand low temperatures and, provided freezing and thawing rates are optimal, cultures of bacteria, algae, yeast and other fungi, and many protozoan species, can be frozen to the temperature of liquid nitrogen and rethawed without undue loss.

The responses to freezing of animal cell suspensions have been extensively studied and we have mentioned previously the pioneer work done on the storage of red blood cells and the development of protective additives such as glycerol. The freezing and preservation of spermatozoa has also been perfected and is now of great economic significance, particularly in cattle breeding by artificial insemination. Human spermatozoa have also been preserved at low temperature with the addition of glycerol and have been used successfully in the clinical treatment of certain types of infertility. However, perhaps even more fascinating is the recent work done on the preservation of early mammalian embryos. It has proved possible to store at −197°C embryos at all stages from the single-celled fertilized ovum to the blastocyst. Such embryos can thus be kept indefinitely and, when required, thawed and transferred into the uterus of a foster mother to continue their development. To date, mouse embryos have been successfully stored for five years and the technique has been used on early embryos from cow, sheep, rabbit and rat.

This is the nearest we have yet got to 'suspended animation' in that a *potential* mammalian individual can be immortalized, but it should be remembered that such early embryos are little more than a clump of simple and effectively identical cells – the problems arise when attempts are made to preserve groups of cells of disparate structure and function (see below).

8.3.2 Freezing organized tissues

Small fragments of skin and corneas can be frozen in the presence of glycerol or dimethyl sulphoxide, stored at low temperature, and successfully transplanted into recipient animals. In man, grafting of the cornea after such low temperature storage is now well established. However, although functional survival after freezing is possible in these tissues and indeed in small fragments of many other tissues, it is safe to say that at the present time there have been no reports of complete resumption of function in any major mammalian organ after total freezing. Many attempts have been made to preserve hearts and kidneys at low temperatures but neither the addition of protective additives nor the manipulation of freezing or thawing rates have been successful in allowing the normal resumption of the heart-beat or kidney function after freezing.

8.3.3 Freezing whole animals

In view of the preceding remarks on the lack of success of attempts to freeze single organs, it is perhaps superfluous to note that attempts to revive whole mammals after complete freezing have been universally unsuccessful. Nevertheless, work on cooling and rewarming mammals has not been entirely unrewarding, for some of what has been learned about the physiology of hypothermia is now finding applications in medicine (p. 77).

The classic work on whole mammal freezing was that of Audrey U.

Smith (described in detail in SMITH, 1961). In these experiments it was found that the body temperature of anaesthetized hamsters could be lowered to between 0°C and −1°C by immersion in a −5°C bath. Under these conditions the heart beat and respiration ceased, the skin temperature fell to −4.5°C and ice crystals were found in the skin and subcutaneous tissues. A layer of ice formed on the outer surface of the abdominal and thoracic viscera and ice was found in the stomach, intestine and bladder. Animals could be maintained in this condition for up to 45 minutes (30–40% of body water frozen) and, provided that continuous artificial respiration was given while they were rewarmed in a 40°C bath, they recovered and suffered no apparent permanent damage. However, if the period of hypothermia was prolonged beyond about 45 minutes so that more than 50% of the body water became frozen, survival was not possible. In reviewing these experiments Smith concludes that 'with the temperatures used, adult golden hamsters would not survive freezing for more than 1 hour at −5°C or after more than 50% of the body water had frozen. . . . It is clear that there is little prospect of storing them in a state of suspended animation in a partially frozen condition at temperatures close to zero, and little hope of reviving them after complete freezing at lower temperatures.' Such results in small animals such as hamsters, where cooling is relatively simple, do not predispose one to think that it will be possible in the foreseeable future to maintain human subjects in a state of suspended animation, for the problems of freezing a much larger mass of tissue are considerably more acute. Notwithstanding the impossibility of freezing and reviving human subjects there are at the present time in the U.S.A. several commercial centres where bodies are stored at the temperature of liquid nitrogen in the optimistic expectation that they will ultimately be revived when medical science is ready with a cure for the disease responsible for their deaths. Perhaps the existence of such establishments can be attributed in part to the many bizarre instances in folk-lore of reanimation after apparent death typified by the case of Roger Dodsworth. The frozen body of this gentleman was found under a pile of snow in Switzerland in 1826 and, after resuscitation, he claimed to have been buried in an avalanche in 1660! There mere fact that such a claim has been documented says much for the appeal of such stories to the public imagination.

8.4   Freeze-drying (lyophilization)

Freeze-drying is a means of dehydrating biological substances in such a way that the removal of water causes minimal damage to them. This is achieved by first freezing the material and then allowing it to rewarm under reduced pressure. With a proper balance between heat input and vapour removal, the ice sublimes and the water is removed in the vapour phase. During the initial freezing, water in the specimen which is not chemically bound separates as pure ice crystals and the subsequent extraction of this water as vapour can be achieved with no shrinkage of the specimen, while the chemical changes normally associated with progressive dehydration as liquid water are minimized.

Freeze-drying has many applications in biology. The preservation of structure and prevention of shrinkage are of great value in the preparation of specimens for histological examination and also in the preservation of relatively large tissue masses, such as pathological specimens, and even whole animals for museum display. The great reduction in chemical changes during dehydration has been exploited in the pharmaceutical industry and in the preparation of cultures of microorganisms. It has also found wider industrial application in the preparation of dehydrated foods, which retain their natural taste during the process, are of small volume relative to the hydrated product, and can be stored at room temperature.

Freeze-drying has, as yet, found relatively little application in the preservation of animal cells in a state that will allow recovery of function. Red blood cells can be preserved by freeze-drying, but the reconstituted cells show a poorer recovery than cells simply frozen with the addition of glycerol. Bovine spermatozoa have been freeze-dryed and rehydrated, and have recovered their motility and been used in successful inseminations. However, different investigators report variable results in the latter case and it is fair to say that recovery of sperm function after freeze-drying has been the exception rather than the rule.

A comparison of simple freezing with freeze-drying as a means of preserving biological material indicates that freezing would be the method of choice if a functional recovery was desired, whereas, if this is not of prime concern, freeze-drying has much to commend it, in particular the fact that freeze-dried material does not require refrigeration during storage.

8.5   Medical applications of hypothermia

Cold has many applications in medicine, ranging from the reduction of fever to the treatment of certain forms of cancer, but space permits only a brief description of two of the more interesting examples.

8.5.1   Cryosurgery

In the treatment of disorders of the brain it is sometimes found that a condition can be alleviated by the deliberate destruction of another part of the brain. Small areas of brain tissue can be destroyed by local heating at the tip of a special electrode positioned under X-ray guidance or by the accurate injection of alcohol. Both these methods have the disadvantage that the damage caused is irreversible, so that if the electrode or needle is misplaced, the treatment may not have the desired effect and may indeed make the patient worse. A technique has recently been developed which allows the destruction of selected areas of the brain by extreme cold (cryosurgery). This method, which involves the localized injection of liquid nitrogen, has the advantage that it is possible to assess the effect of damage in a particular area before the effects become irreversible. This technique has found particular application in the treatment of Parkinson's disease (paralysis agitans) where the tremor and paralysis can often be cured by localized destruction of part of the basal ganglia. This operation is performed under local anaesthesia so the surgeon can see the result of cooling different parts of the basal ganglia before finally destroying permanently the appropriate region.

8.5.2 Hypothermia

The other field of medicine which has advanced dramatically through the use of cold is heart surgery. The use of hypothermia, together with a mechanical means of perfusing the body with oxygenated blood (heart–lung machine), allows the heart to be stopped and drained of blood for relatively long periods of time. Under these circumstances, surgeons are able to perform delicate and often prolonged repairs to internal parts of the heart (open-heart surgery). The absence of blood, the fact that the heart is not beating and the relatively long time available permit procedures which would be technically impossible in the beating blood-filled heart.

When the circulation is stopped at normal body temperatures, permanent brain damage will ensue within 3–5 minutes because the brain cells are being deprived of oxygen. However, if the brain is cooled, its metabolism slows and its oxygen requirement is greatly reduced. For this reason, it is possible to stop the heart for longer periods during hypothermia. Indeed, at a body temperature of 15°C, the heart can be safely stopped for up to one hour so that open-heart surgery which can be accomplished in less than this time is usually performed under hypothermia alone. For periods longer than one hour, however, hypothermia must be accompanied by mechanical perfusion of oxygenated blood to the brain and the rest of the body.

Surgical hypothermia can be induced by cooling the body surface or by cooling the blood in an external heat-exchanger. In surface cooling the body is cooled by placing the anaesthetized patient in an ice bath or refrigerated blankets. The shivering response is abolished by a combination of muscle-relaxant drugs and the general anaesthetic, and vasodilation in the skin is produced by other drugs. In the absence of shivering, and with a brisk skin blood flow, the skin acts as a very efficient heat-exchanger and the body temperature falls relatively quickly. At the end of the operation the patient is rewarmed by immersion in a bath at 40°C. Surface cooling is rather cumbersome, however, and has now generally been replaced by techniques where the blood is cooled in an external heat-exchanger. There are many ways in which this can be accomplished, one of which is illustrated in Fig. 8–1. In this technique blood returning from the lungs is drained from the left atrium into a

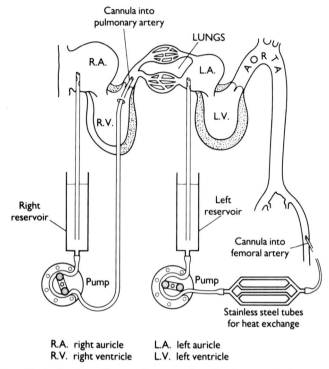

Cannula into
pulmonary artery

LUNGS

R.A.

L.A.

AORTA

R.V.

L.V.

Right
reservoir

Left
reservoir

Cannula into
femoral artery

Pump

Pump

Stainless steel tubes
for heat exchange

R.A. right auricle   L.A. left auricle
R.V. right ventricle   L.V. left ventricle

Fig. 8–1 Method of cooling and rewarming the human body to provide profound hypothermia for cardiac surgery. (From DREW, C. E. (1961). *Brit. Med. Bulletin*, **17**, 38.)

reservoir, from which a pump taking the place of the left ventricle drives it via a heat-exchanger into the arterial system: the blood is cooled during passage through the heat-exchanger which comprises stainless steel pipes immersed in iced water. As cooling proceeds, the heart begins to fail and the perfusion of the lungs is taken over by a second pump replacing the right ventricle. Cooling continues until the body temperature reaches 15°C, at which point the pumps are stopped and surgery begins. The sequence is reversed during rewarming when the heat-exchanger bath is filled with water at 40°C. This technique is more convenient than surface cooling and much more rapid; the cooling and rewarming cycles can usually be completed within about 30 minutes.

Conclusion

We have seen in this book how homeothermic animals have evolved means of regulating their internal temperature in such a way that they are able to function efficiently throughout a wide range of environmental temperatures. Poikilotherms, on the other hand, have not escaped from the thermal limitations of their environment and are, in consequence, severely limited in their ecological potential.

However, homeothermy cannot simply be dismissed as an evolutionary trick calculated to allow the exploitation of thermally difficult habitats, because it has perhaps even greater significance as a prerequisite for physiological sophistication. For without homeothermy, the development and functioning of intricate biological mechanisms such as the human brain would be inconceivable.

Appendix: some suggestions for practical work

A.1 Q_{10} and the Arrhenius Relation

The isolated frog's heart provides a convenient preparation for the demonstration of the effect of temperature on activity.

Pith the frog in the usual way and remove a large piece of skin from the ventral surface over the pectoral girdle, and then use fine scissors to cut through the insertion of the abdominal muscles into the xiphisternum. Provided that the incision is kept close to the xiphisternum it is not necessary to tie off the abdominal vein. Raise the xiphisternum and, keeping the points well up to avoid damaging the heart, use coarse scissors to cut through the pectoral girdle on both sides. Remove the central portion.

The heart should now be visible, beating inside the pericardium. Determine the rate of beating and then note the frog's temperature (the latter can be done by inserting the bulb of a mercury thermometer amongst the abdominal viscera). The heart can now be excised. Insert a small bent pin attached to a fine thread into the apex of the ventricle and use this gently to lift the heart. Pass a fine thread about 20 cm long under one branch of the aorta and tie it firmly, arranging the knot in the centre of the thread. Then lift carefully with the two threads now attached to the heart, cut through the remaining arterial and venous connections and free the heart (N.B. it is most important to avoid damage to the sinus venosus or the atria). Remove the pin from the ventricle and submerge the heart until required in a petri dish of Frog Ringer's solution at 20°C.

The apparatus, see Fig. A–1, comprises a large boiling tube, sealed by a rubber bung and almost filled with Frog Ringer's solution. Oxygen or compressed air is bubbled slowly into the solution from a long tube passing through the bung and extending almost to the bottom of the tube. The gas oxygenates the solution and also aids mixing and thus more rapid thermal equilibration with the water bath. The temperature of the Ringer's solution is monitored by a short thermometer fastened to the oxygen tube by elastic bands. With the apparatus initially at 20°C attach the heart to the oxygen tube by the thread as shown in Fig. A–1.

Determine the rate of beating at this temperature (it is helpful to turn off the oxygen supply while actually counting the heart beats). Freshly excised hearts may beat more rapidly than before removal and the rate then declines to a stable plateau value. When the rate is relatively stabilized, alter the temperature of the water bath within the range 5–25°C and measure the change in heart rate. Plot the results on linear graph paper (heart rate, abscissa v centigrade temperature, ordinate) and

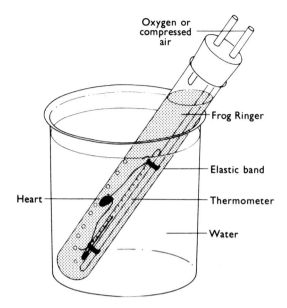

Oxygen or
compressed
air

Frog Ringer

Elastic band

Thermometer

Water

Heart

Fig. A–1 Apparatus for measuring the effect of temperature on the isolated frog heart.

obtain a value for the Q_{10}. Mark on this graph the value you obtained earlier with the heart still in the frog. You can also plot the results on semi-log graph paper (log heart rate, abscissa v reciprocal of the absolute temperature, ordinate) and obtain an Arrhenius plot (see p. 2). (A fuller discussion of the implications of this experiment and suggestions for further work can be found in the classical paper of BARCROFT, J. and IZQUIERDO, J. J. (1930). *J. Physiol.*, **71**, 145–55.) When you have obtained all the data you require from the heart, allow the temperature to increase slowly to values above 25°C and note the behaviour of the heart.

A.2   The effect of cold on flying insects

Obtain a transparent-sided container, the largest that will fit into your refrigerator (a small aquarium tank or a 2 l beaker will suffice). Cover the container with a sheet of cardboard and suspend from this a small thermometer, making sure that it does not touch the walls of the container. Introduce a few individuals of any suitable-sized flying insect (e.g. house flies, mosquitoes, midges) and place the container in the refrigerator. When it has cooled sufficiently to render the insects immobile, remove it and allow it to warm slowly to room temperature. Note the temperature at which spontaneous flight first takes place and record your observations of the insects' behaviour during the pre-flight period. Repeat the experiment and see if you can provoke flight at lower temperatures by disturbing the insects. Compare the behaviour of different types of insects and see if any differences can be related to their normal patterns of life (certain butterflies and moths provide striking contrasts).

A.3 Thermal gradients and temperature preferenda

Take a block of metal, preferably copper or brass, about 5 cm wide and 3 cm thick and at least 50 cm long. Cover the central observation area in with a transparent chamber (perspex) and fasten a strip of marked graph paper onto the edge of the metal below the cover (Fig. A–2). Establish a

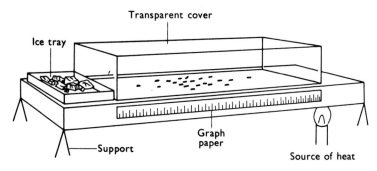

Fig. A–2 Temperature gradient chamber.

suitable thermal gradient (5–45°C) along the strip by placing an ice container at one end and gently warming the other end. Calibrate the gradient within the observation chamber by placing the bulb of a mercury thermometer against the metal at intervals. Put the experimental animals (e.g. house fly or meal-worm larvae) into the chamber, scattering them at random along its length. Leave the apparatus under uniform lighting conditions and plot the distribution of individual animals at intervals. Devise experiments to see if previous acclimatization to heat or cold alters the preferred temperature.

A.4 Experiments on man

A.4.1 Oral temperature

The temperature in the oral cavity, measured with a clinical thermometer, provides a convenient approximate index of core temperature.

Measure your oral temperature and determine the effect of drinking either very hot liquid or iced water on oral temperature. How long do the effects persist and are they true reflections of changes in *core* temperature?

Investigate the effect of exercise on oral temperature. Devise a standardized exercise system (e.g. knee bends at 60/min or steppping rapidly on and off a box) and perform this for various lengths of time. Does the environmental temperature influence your results?

A.4.2   Response to cold

Man is homeothermic, so that a moderate fall in environmental temperature should not lower core temperature significantly. However, skin temperatures in different parts of the body will fall appreciably in a cold environment. Record the oral temperature and the skin temperature at various sites in a subject during cooling. The latter may be measured by gently pressing the tip of a mercury thermometer on to the skin having previously covered the upper half of the bulb with a small cork. Cooling can be achieved if the subject, wearing the minimum of clothing, lies on a wooden bench or table in an environmental temperature below 20°C. Record the subjective impressions of the person (without allowing him to know the recorded skin temperatures) and observe the changes in the appearance of the skin (colour, horripilation). Note also the oral and skin temperatures at the onset of shivering. Repeat the experiment immediately after the subject has had a hot bath or has indulged in strenuous exercise. Consider your results in the light of Benzinger's experiments (see p. 35).

Further Reading

AVERY, R. A. (1979). *Lizards – A Study in Thermoregulation*. Studies in Biology no. 109. Edward Arnold, London.

BARRINGTON, E. J. W. (1968). *The Chemical Basis of Physiological Regulation*. Scott, Foresman and Co., Illinois.

BENZINGER, T. H. (1961). The human thermostat. *Scient. Am.*, **204**, 134–47.

BENZINGER, T. H. (1969). Heat regulation: homeostasis of central temperature in man. *Physiol. Rev.*, **49**, 671–759.

BLIGH, J. (1966). The thermosensitivity of the hypothalamus and thermoregulation in mammals. *Biol. Rev.*, **41**, 317–67.

BLIGH, J. (1973). *Temperature Regulation in Mammals and Other Vertebrates*. North–Holland Publishing Company, Amsterdam, London.

BOGERT, C. M. (1959). How reptiles regulate their body temperature. *Scient. Am.*, **200**, 105–20.

CHAPMAN, R. F. (1969). *The Insects*. The English Universities Press, London.

DAVSON, H. and SEGAL, M. B. (1975). *Introduction to Physiology*, Vol. I. Academic Press, London, New York.

GORDON, M. S., BARTHOLEMEW, G. A., GRINNELL, A. D., JØRGENSEN, C. B. and WHITE, F. N. (1968). *Animal Function: Principles and Adaptations*. Macmillan, New York and London.

IRVING, L. (1966). Adaptations to cold. *Scient. Am.*, **214**, 94–101.

KAYSER, C. (1961). *The Physiology of Natural Hibernation*. Pergamon, Oxford.

PROSSER, C. L. and BROWN, F. A. (1973). *Comparative Animal Physiology* (3rd edn). W. B. Saunders Company, Philadelphia and London.

RUCH, T. C. and PATTON, H. D. (1973). *Physiology and Biophysics*, Vol. III. W. B. Saunders Company, Philadelphia and London.

SCHMIDT-NIELSEN, K. (1964). *Desert Animals*. The Clarendon Press, Oxford.

SMITH, AUDREY U. (1961). *Biological Effects of Freezing and Supercooling*. Edward Arnold, London.

SWAN, H. (1974). *Thermoregulation and Bioenergetics*. Elsevier, New York, London.